Geology

by James Geikie

PREFACE.

The vital importance of diffusing some knowledge of the leading principles of Science among all classes of society, is becoming daily more widely and deeply felt; and to meet and promote this important movement, W. & R. CHAMBERS have resolved on issuing the present Series of ELEMENTARY SCIENCE MANUALS. The Editors believe that they enjoy special facilities for the successful execution of such an undertaking, owing to their long experience--now extending over a period of forty years--in the work of popular education, as well as to their having the co-operation of writers specially qualified to treat the several subjects. In particular, they are happy in having the editorial assistance of ANDREW FINDLATER, LL.D., to whose labours they were so much indebted in the work of editing and preparing Chamber's Encyclopaedia.

The Manuals of this series are intended to serve two somewhat different purposes:

1. They are designed, in the first place, for SELF-INSTRUCTION, and will present, in a form suitable for private study, the main subjects entering into an enlightened education; so that young persons in earnest about self-culture may be able to master them for themselves.

2. The other purpose of the Manuals is, to serve as TEXT-BOOKS IN SCHOOLS. The mode of treatment naturally adopted in what is to be studied without a teacher, so far from being a drawback in a school-manual, will, it is believed, be a positive advantage. Instead of a number of abrupt statements being presented, to be taken on trust and learned, as has been the usual method in school-teaching; the subject is made, as far as possible, to unfold itself gradually, as if the pupil were discovering the principles himself, the chief function of the book being, to bring the materials before him, and to guide him by the shortest road to the discovery. This is now acknowledged to be the only profitable method of acquiring knowledge, whether as regards self-instruction or learning at school.

For simplification in teaching, the subject has been divided into sub-sections or articles, which are numbered continuously; and a series of Questions, in corresponding divisions, has been appended. These Questions, while they will enable the private student to test for himself how far he has mastered the several parts of the subject as he proceeds, will serve the teacher of a class as specimens of the more detailed and varied examination to which he should subject his pupils.

NOTE BY THE AUTHOR.

In the present Manual of GEOLOGY it has been the aim of the author rather to indicate the methods of geological inquiry and reasoning, than to present the learner with a tedious summary of results. Attention has therefore been directed chiefly to the physical branches of the science--Palaeontology and Historical Geology, which are very large subjects of themselves, having been only lightly touched upon. The student who has attained to a fair knowledge of the scope and bearing of Physical Geology, should have little difficulty in subsequently tackling those manuals in which the results obtained by geological investigation are specially treated of.

CONTENTS.

PAGE INTRODUCTORY

CLASSIFICATION OF ROCKS

MINERALOGY 12 ROCK-FORMING MINERALS

PETROLOGY-- MECHANICALLY FORMED ROCKS 17 CHEMICALLY FORMED ROCKS ORGANICALLY DERIVED ROCKS 20 METAMORPHIC ROCKS IGNEOUS ROCKS STRUCTURE AND ARRANGEMENT OF ROCK-MASSES-- Stratification, &c.; Mud-cracks and Rain-prints; Succession of Strata; Extent of Beds; Sequence of Beds--Joints;

Cleavage; Foliation; Concretions; Inclination of Strata; Contemporaneous Erosion; Unconformability; Overlap; Faults; Mode of Occurrence of Metamorphic and Igneous Rocks; Mineral Veins 26-46

DYNAMICAL GEOLOGY-- THE ATMOSPHERE AS A GEOLOGICAL AGENT OF CHANGE 46 WATER AS A GEOLOGICAL AGENT OF CHANGE 48 GEOLOGICAL ACTION OF PLANTS AND ANIMALS SUBTERRANEAN FORCES   METAMORPHISM

PHYSIOGRAPHY

PALAEONTOLOGY

HISTORICAL GEOLOGY

QUESTIONS

INTRODUCTORY.

1. Definition.--Geology is the science of the origin and development of the structure of the earth. It treats of the nature and mode of formation of the various materials of which the earth's crust is composed; it seeks to discover what mutations of land and water, and what changes of climate, have supervened during the past; it endeavours to trace the history of the multitudinous tribes of plants and animals which have successively tenanted our globe. In a word, Geology is the Physical Geography of past ages.

2. Rocks.--Every one knows that the crust of the earth is composed of very various substances, some of which are hard and crystalline in texture, like granite; others less indurated and non-crystalline, such as sandstone, chalk, shale, &c.; while yet others are more or less soft and incoherent masses, as gravel, sand, clay, peat, &c. Now, all these heterogeneous materials, whether they be hard or soft, compact or loose, granular or crystalline, are termed rocks. Blowing sand-dunes, alluvial silt and sand, and even peat, are, geologically speaking, rocks, just as much as basalt or any indurated building-stone. The variety of rocks is very great, but we do not study these long before we become aware that many kinds which present numerous contrasts in detail, yet possess certain characters in common. And this not only groups these diverse species together, but serves also to distinguish them from other species of rock, which in like manner are characterised by the presence of some prevalent generic feature or features.

Classification of Rocks.--All the rocks that we know of are thus capable of being arranged under five classes, as follows:

I. MECHANICALLY FORMED. II. CHEMICALLY FORMED. III. ORGANICALLY DERIVED. IV. METAMORPHIC. V. IGNEOUS.

3. The MECHANICALLY FORMED class comprises a considerable variety of rocks, all of which, however, come under only two subdivisions--namely, Sedimentary, and Eolian or Aerial, the former being by far the more important.

Of the Sedimentary group, there are three rocks which may be taken as typical and representative--namely, conglomerate or puddingstone, sandstone, and shale. A short examination of the nature of these will sufficiently explain why they come to be grouped together under one head. Conglomerate consists of a mass of various-sized rounded stones cemented together; each stone has been well rubbed, and rolled, and rounded. It is quite obvious that the now solid rock must at one time have existed in a loose and unconsolidated state, like gravel and shingle. Nor can we resist the conclusion that the stones were at one time rolled about by the action of water--that being the only mode in which gravel-stones are shaped. Again, when we have an opportunity of examining any considerable vertical thickness of conglomerate, we shall frequently observe that the stones are arranged more or less definitely along certain lines. These, there can be no question, are lines of deposition--the rounded stones have evidently not been formed and accumulated all at once, but piled up gradually, layer upon layer. And since there is no force in nature, that we know of, save water in motion, that could so round and smooth stones, and spread them out in successive layers or beds, we may now amplify our definition of conglomerate, and describe it as a compacted mass of stones which have been more or less rounded, and arranged in more or less distinct layers or beds, by the action of water.

4. Sandstone may at the outset be described as a granular non-crystalline rock. This rock shews every degree of coarseness, from a mass in which the constituent grains are nearly as large as turnip-seed, down to a stone so fine in the grain that we need a lens to discover what the particles are of which it is composed. When these latter are examined, they are found to exhibit marks of attrition, just like the stones of a conglomerate. Sharp edges have been worn off, and the grains rounded and rubbed; and whereas lines of deposition are often obscure, and of infrequent occurrence in conglomerate--in sandstone, on the contrary, they are usually well marked and often abundant. We can hardly doubt, therefore, that sandstone has also had an aqueous origin, or in other words, that it has been formed and accumulated by the force of water in motion. In short, sandstone is merely compacted sand.

5. If it be easy to read the origin of conglomerate and sand in the external character of their ingredients, and the mode in which these have been arranged, we shall find it not less easy to discover the origin of shale. Shale is, like sandstone, a granular non-crystalline rock. The particles of which it is built up are usually too small to be distinguished without the aid of a lens, but when put under a sufficient magnifying power, they exhibit evident marks of attrition. In structure it differs widely from sandstone. In the latter rock the layers of deposition, though frequently numerous, are yet separated from each other by some considerable distance, it may be by a few inches or by many yards. But in shale the layers are so thin that we may split the rock into laminae or plates. Now we know that many sedimentary materials of recent origin, such as the silt of lakes, rivers, and estuaries, although when newly dug into they appear to be more or less homogeneous, and shew but few lines of deposition, yet when exposed to the action of the atmosphere and dried, they very often split up into layers exhibiting division planes as minute as any observable in shale. There is no reason to doubt, therefore, that shale is merely compacted silt and mud--the sediment deposited by water. It becomes evident, therefore, that conglomerate, sandstone, and shale are terms of one series. They are all equally sedimentary deposits, and thus, if we slightly modify our definition of conglomerate, we shall have a definition which will include the three rocks we have been considering. For they may all be described as granular non-crystalline rocks, the constituent ingredients of which have been more or less rounded, and arranged in more or less distinct layers, by the action of water.

6. The Eolian or Aerial group of rocks embraces all natural accumulations of organic or inorganic materials, which have been formed upon the land. The group is typically represented by debris, such as gathers on hill-slopes and at the base of cliffs, by the sand-hills of deserts and maritime districts, and by soil. All these accumulations owe their origin to atmospheric agencies, as will be more particularly described in the sequel. As the Sedimentary and Eolian rocks are the results of the mechanical action of water and the atmosphere, they are fitly arranged under one great class--the MECHANICALLY FORMED ROCKS.

7. CHEMICALLY FORMED ROCKS constitute another well-marked class, of which we may take rock-salt as a typical example. This rock has evidently been deposited in water, but not in the manner of a sedimentary bed. It is not built up of water-worn particles which have been rolled about and accumulated layer upon layer, but has been slowly precipitated during the gradual evaporation of water in which it was previously held in solution. Its formation is therefore a chemical process. Various other rocks come under the same category, as we shall afterwards point out.

8. The ORGANICALLY DERIVED class comprises a number of the most important and useful rock-masses. Chalk may be selected as a typical example. Even a slight examination shews that this rock differs widely from any of those mentioned above. Conglomerate, sandstone, shale, &c. are built up of pebbles, particles, grains, &c. of various inorganic materials. But chalk, when looked at under the microscope, betrays an organic origin. It consists, chiefly, of the hard calcareous parts of animal organisms, and is more or less abundantly stocked with the remains of corals, shells, crustaceans, &c. in every degree of preservation; indeed, so abundant are these relics, that they go to form a great proportion of the rock. Coal is another familiar example of an organically derived rock, since it consists entirely of vegetable remains.

9. The METAMORPHIC class, as the name implies, embraces all those rocks which have undergone some decided change since the time of their formation. This change generally consists in a re-arrangement of their constituent elements, and has frequently resulted in giving a crystalline texture to the rocks affected. Hence certain sedimentary deposits like sandstone and shale have been changed from granular into crystalline rocks, and the like has happened to beds of limestone and chalk. Mica-schist, gneiss, and saccharoid marble are typical of this class.

10. The IGNEOUS rocks are those which owe their origin to the action of the internal forces of the earth's crust. Most of them have been in a state of fusion, and betray their origin by their crystalline and sometimes glassy texture, and

also, as we shall see in another section, by the mode of their occurrence. Lava, basalt, and obsidian are characteristic types of this group of igneous rocks. Another group embraces a large variety of igneous rocks which are non-crystalline, and vary in texture from fine-grained, almost compact, bedded masses, like certain varieties of tuff, up to coarse, irregular accumulations of angular stones imbedded in a fine-grained or gritty matrix, like volcanic breccia and volcanic agglomerate.

MINERALOGY.

11. Having learned that all the rocks met with at the surface of the earth's crust are capable of being arranged under a few classes, we have now to investigate the matter more in detail. It will be observed that the classification adopted above is based chiefly upon the external characters of the constituent ingredients of the rocks, and the mode in which these particles have been collected. In some rocks the component materials are crystalline, in others they are rounded and worn; in one case they have been brought together by precipitation from an aqueous solution, or they have crystallised out from a mass of once molten matter; in another case their collection and intimate association is due to the mechanical action of the atmosphere or of water, or to the agency of the organic forces. We have next to inquire what is the nature of those crystals and particles which are the ingredients of the rocks? The answer to this question properly belongs to the science of mineralogy, with which, however, the geologist must necessarily make some acquaintance.

12. Granite--its composition.--It will tend to simplify matters if we begin our inquiry by selecting for examination some familiar rock, such as granite. This rock, as one sees at a glance, is crystalline, nor is it difficult to perceive that three separate kinds of ingredients go to compose it. One of these we shall observe is a gray, or it may be, clear glassy-looking substance, which is hard, and will not scratch with a knife; another is of a pink, red, gray, or sometimes even pale green colour, and scratches with difficulty; while the third shews a glistering metallic lustre, and is generally of a brownish or black colour. It scratches easily with the knife, and can be split up into flakes of extreme

thinness. If the granite be one of the coarse-grained varieties, we shall notice that these three ingredients have each more or less definite crystalline forms; so that they are not distinguished by colour and hardness alone. The metallic-looking substance is mica; the hard gray, or glassy and unscratchable ingredient is quartz; and the remaining material is felspar. The mineralogist's analysis of granite ends here. But there is still much to be learned about quartz, felspar, and mica; for, as the chemist will tell us, these are not 'elementary substances.' Quartz is a compound, consisting of two elements, one of which is a non-metallic body (silicon), and the other an invisible gas (oxygen). Felspar[A] is a still more complex compound, being made up of two metals (potassium, aluminium) and one non-metallic body (silicon), each of which is united to an invisible gas (oxygen). Mica, again, contains no fewer than four metals (potassium, magnesium, iron, calcium) and one non-metallic body (silicon), each of which is in like manner chemically united to its share of oxygen. Thus the rock-forming substances, quartz, felspar, and mica, have each a definite chemical composition.

13. Minerals.--Now, any inorganic substance which has a definite chemical composition, and crystallises in a definite crystalline or geometric form, is termed a mineral. Having once discovered that quartz is composed of silicon and oxygen--that is, silica--and that the faces of its crystals are arranged in a certain definite order, we may be quite sure that any mineral which has not this composition and form cannot be quartz. And so on with mica and felspar, and every other mineral. The study of the geometric forms assumed by minerals (crystallography) forms a department of the science of mineralogy. But, in the great majority of cases, the mineral ingredients of the rocks are either so small individually, or so broken, and rounded, and altered, that crystallography gives comparatively little aid to the practical geologist in the field. He has, therefore, recourse to other tests for the determination of the mineral constituents of rocks. Many of these tests, however, can only be applied by those who have had long experience. The simplest and easiest way for the student to begin is to examine the forms and appearance of the more common minerals in some collection, and thereafter to accustom his eye to the aspect presented by the same minerals when they are associated together in rocks, of which illustrative

specimens are now to be met with in most museums. The microscope is largely employed by geologists for determining the mineralogical composition of certain rocks; and, indeed, many rocks can hardly be said to be thoroughly known until they have been sliced and examined under the microscope, and analysed by the chemist. But with a vast number such minute examination is not required, the eye after some practice being able to detect all that is needful to be known.

[A] There are various kinds of felspar; the one referred to above is orthoclase, or potash-felspar.

ROCK-FORMING MINERALS.

14. Nearly all the minerals we know of contain oxygen as a necessary ingredient, there being only a very few minerals in which that gas does not occur in chemical union with other elements. Three of these minerals, sulphur, the diamond, and graphite, consist of simple substances, and are of great commercial importance, but none of them is of so frequent occurrence, as a rock constituent, as the minerals presently to be described. Sulphur occurs sometimes in thin beds, but more frequently in small nests and nodules, &c. in other rocks, or in joints, and fissures, and veins. It is frequently found in volcanic districts. The diamond, which consists of pure carbon, is generally met with in alluvial deposits, but sometimes, also, in a curious flexible sandstone, called itacolumite. Graphite is another form of carbon. It occurs both in a crystalline and amorphous form, the latter, or non-crystalline kind, being extensively used for lead-pencils. Rock-salt is a chloride of sodium, and appears sometimes in masses of a hundred feet and more in thickness. Another mineral which contains no oxygen is the well-known fluor-spar. It occurs chiefly in veins, and is often associated with ores. With these, and a few other exceptions, all the minerals hitherto discovered contain oxygen as an essential element; and so large is the proportion of this gas which enters into union with other elements to constitute the various minerals of which the rocks are composed, that it forms at least one-half of all the ponderable matter near the earth's surface. When the student learns that there are probably no fewer than

six or seven hundred different minerals, he will understand how impossible it is to do more in a short geological treatise than point out a few of the most commonly occurring ones. And, indeed, a knowledge of the chief rock-forming minerals, which are few in number, is all that is absolutely requisite for geological purposes. Some of these we accordingly proceed to name.[B]

[B] It is needless to describe the minerals minutely here. The student can only learn to distinguish the different species by carefully examining actual specimens.

15. Quartz.--This mineral has already been partially described. It is the most abundant of all the rock-forming minerals, and occurs in three forms: (1) crystallised quartz or rock crystal; (2) chalcedony, both of which are composed of silica--that is, silicon and oxygen; and (3) hydrated quartz--that is, silica with the addition of water.

Hematite.--This is an oxide of iron. It occurs in mammillary rounded masses, with a fibrous structure, and a dull metallic lustre. Magnetite or magnetic iron ore, specular iron, and limonite are also oxides of iron. Hematite shews a red streak when scratched with a knife, which distinguishes it from magnetite.

Iron pyrites.--This is a sulphide of iron of very common occurrence. Its crystalline form is cubical. When broken, it emits a sulphurous smell. The brass-yellow coloured cubes so often seen in roofing-slates are familiar examples of the mode of its occurrence. But it is also frequently found in masses having a crystalline surface.

16. SULPHATES.--Only two sulphates may be noticed--namely, gypsum, which is a sulphate of lime, with its varieties, selenite, satin-spar, and alabaster; and barytes, a sulphate of baryta. Barytes scratches easily with the knife, and from its great specific gravity is often called heavy-spar. Gypsum is softer than barytes.

CARBONATES.--Two of these only need be mentioned: calcite or calc-spar, a carbonate of lime, which scratches with the knife, and effervesces readily with dilute hydrochloric acid; and arragonite, also a carbonate of lime, but denser than calcite.

SILICATES.--These are by far the most abundantly occurring minerals. The species are also exceedingly numerous, but we may note here only a few of the more important. They are composed of silica and various bases, such as lime, potash, magnesia, soda, alumina, &c. Augite or pyroxene is a black or greenish-black mineral, found, either as crystals, which are generally small, or as rounded grains and angular fragments, in basaltic and volcanic rocks. It never occurs in granite rocks. It is brittle, and has a vitreous or resinous lustre. There are a number of varieties or sub-species of augite. Hornblende, like augite, also includes a great many minerals. When the crystals are small, it is often difficult to distinguish hornblende from augite. Common hornblende occurs crystallised or massive, and is dark green or black, with a vitreous lustre. It is generally sub-translucent. It usually crystallises in igneous rocks which contain much quartz or silica; while augite, on the other hand, crystallises in igneous rocks which are of a more basic character--that is to say, rocks in which silica is not so abundantly present. Felspar is a generic term which embraces a number of species, such as orthoclase or potash-felspar, albite or soda-felspar, and anorthite or lime-felspar. Orthoclase is white, red or pink, and gray. It is one of the ordinary constituents of granite, and enters into the composition of many rocks. Albite is usually white. It often occurs as a constituent of granite, not unfrequently being associated in the same rock with pink felspar or orthoclase. In syenite and greenstone it occurs more commonly than orthoclase. Anorthite occurs in white translucent or transparent crystals. It is not so common a constituent of rocks as either of the other felspars just referred to. Mica: this term includes several minerals, which all agree in being highly cleavable into thin elastic flakes or laminae, which have a glistening metallic lustre. Mica is one of the common constituents of granite. Talc is a silvery white, grayish, pale or dark-green coloured mineral, with a pearly lustre. It splits readily into thin flakes, which are flexible, but not elastic, and may be readily scratched with the nail. It is unctuous and greasy to the touch. It occurs in beds (talc-slate), and is often met with in districts occupied by metamorphic crystalline rocks. Serpentine is generally of a green colour, but brown, red, and variously mottled varieties occur. It has a dull lustre, and is soft, and easily cut; it is tough, however, and takes on a good polish. It forms rock-masses in some places. The finer varieties are called noble serpentine. Chlorite is another soft, easily scratched mineral, generally of a dark-green colour. It has a pearly lustre. Sometimes it occurs in beds (chlorite-slate), and is often found coating the walls of fissures in certain rocks. It has a somewhat greasy feel. The three last-mentioned minerals--talc, serpentine, and chlorite--are all silicates of magnesia. Zeolites is a term which comprises a number of minerals of varying chemical composition, all of which tend to form a jelly when treated with acids. When heated by the blow-pipe they bubble up, owing to the escape of water; hence their name zeolites, from zeo, I boil, and lithos, a stone. The zeolites occur very commonly in cavities in igneous rocks, and also in mineral veins.

Having now mentioned the chief rock-forming minerals, we proceed to a brief description of some of the more typical representatives of the five great classes of rocks referred to at page 8.

PETROLOGY.[C]

[C] Petros, a rock, and logos, a discourse. Some geologists restrict this term to the study of the structure and arrangement of rock-masses, and apply the term lithology (lithos, a stone, and logos, a discourse) to the study of the mineralogical composition of rocks.

MECHANICALLY FORMED ROCKS.

17. (A.) SEDIMENTARY CLASS.--Three of the most commonly occurring rocks of this class have already been described, but a few details are added here.

Conglomerate.--This is a consolidated mass of more or less water-worn and rounded stones. These stones may be of any size. When they are very large, the rock is called a coarse conglomerate; the finer varieties, in which the stones are small, are known as pebbly conglomerates. The ingredients of a conglomerate may consist of any kind of rock, or of a mixture of many different kinds. When they consist entirely of quartz, the rock becomes quartzose. The finer-grained conglomerates usually shew lines of deposition or bedding, but in some of the coarser sorts it is often difficult to detect any kind of arrangement. The stones are usually imbedded in a matrix of quartzose grit and sand, but sometimes this is very scanty. When the nature of the material which binds the stones together is very well marked, the rock becomes ferruginous, calcareous, arenaceous, or argillaceous, according as the binding or cementing material is iron, lime, sand, or clay. Breccia is a rock in which the included fragments are angular.

18. Sandstone is, as already remarked, merely consolidated sand. The coarser varieties, in which the grains are as large and larger than turnip-seeds, are termed grit. From these coarse varieties the rock passes insensibly, in one direction, into a fine or pebbly conglomerate, and in another into a rock, so fine-grained that a lens is needed to distinguish the component particles. Quartz is the prevailing ingredient--sometimes clear, at other times white. Frequently, however, the grains are coated with an oxide of iron, which gives the resulting rock a red colour. The other colours assumed by sandstone--such as yellow, brown, green, &c.--are also in like manner due to the presence of some compound of iron. When mica or felspar occurs plentifully, we have, in the one case, micaceous sandstone, and in the other felspathic sandstone. A sandstone in which the grains are cemented by carbonate of lime is said to be calcareous. Freestone is a sandstone which can be worked freely in any direction. In most sandstones, the lines of bedding are distinct; when they are so numerous as to render the rock fissile, the sandstone is said to be shaly.

Shale is a more or less indurated fissile or laminated clay. When the rock becomes coarse by the admixture of sand, it gradually passes into a shaly sandstone. There are many other varieties of clay-rocks--such as fire-clay, pipe-clay, marl, loam, &c.--which are sufficiently familiar.

19. (B.) EOLIAN or AERIAL CLASS.--Blown-sand is found at many places on sea-coasts. It generally forms smooth rounded hummocks, which are sometimes arranged in long lines parallel to the trend of the coast, as, for example, in the Tents Moor, near St Andrews. The sand-hills of deserts also belong to this class.

Debris is the loose angular rubbish which collects at the base of cliffs, on hill-tops, and hill-slopes. Immense accumulations of it occur in lofty mountainous districts and in arctic regions. In Nova Zembla, for example, the solid rock of the country is almost concealed beneath a thick covering of debris. But the various kinds of debris will be more particularly described further on.

Soil.--An account of this can hardly be given without entering into the theory of its origin, and therefore we reserve its consideration for the present.

CHEMICALLY FORMED ROCKS.

20. Stalactites and stalagmites are carbonates of lime. They vary in colour, being white, or yellow, or brown. Stalactites are usually found adhering to the roofs of limestone caverns, &c., or depending from limestone rocks; stalagmites, on the other hand, commonly occur on the floors of limestone caverns, where they often attain a thickness of many feet.

Siliceous sinter is silica with the addition of water--in other words, a hydrated quartz. It is not a very abundant rock, and is found chiefly in volcanic countries.

Rock-salt has already been described. It occurs either as thin beds, or in the form of thick cake-like masses, often reaching ninety or one hundred feet in thickness. It is rudely crystalline in texture, and is usually discoloured brown and red with various impurities.

ORGANICALLY DERIVED ROCKS.

21. Limestone consists of carbonate of lime, but usually contains some impurities. The varieties of this rock are numerous; some of them are as follows: Chalk; oolite, a rock built up of little spheroidal concretions, whence its name, egg or roe stone (the coarser oolites are called pisolite, or pea-stone); lacustrine limestone, &c. When much silica is diffused through the rock, we have a siliceous limestone; the presence of clay and of carbonaceous matter gives us argillaceous and carbonaceous limestones. Cornstone is a limestone containing a large quantity of arenaceous matter or sand. Many limestones are distinguished by the different kinds of organic remains which they yield. Thus, we have muschelkalk or shell-limestone, nummulitic, crinoidal, &c. limestone. The crystalline limestones, such as statuary marble, are metamorphosed limestones. Not a few limestones are chemically formed rocks, and many, also, are partly of chemical and partly of organic origin, so that no hard and fast line can be drawn between these two classes of rock.

Dolomite, or magnesian limestone.--This is a compound of carbonate of lime and carbonate of magnesia. Its colour is usually yellow, or yellowish brown, but gray and black varieties are sometimes met with. It is generally fine-grained, with a crystalline texture, and pearly lustre. It effervesces less freely with acids than pure limestone. In many cases dolomite is merely a metamorphosed limestone.

22. Coal is composed of vegetable matter, but usually contains a greater or less percentage of impurities. The varieties of this substance are very numerous, and differ from each other principally in regard to their bituminous or non-bituminous character. Coal is bituminous or non-bituminous according as it is less or more highly mineralised. Bitumen results from the decomposition of vegetable matter; but, when the mineralising process (to which the formation of coal is due) has proceeded far enough, the vegetable matter gradually loses its bituminous character, and the result is a non-bituminous coal. Varieties of coal are the following: Lignite or brown coal; caking coal; cannel, parrot, or gas coal; splint coal; cherry or soft coal; anthracite or blind coal, so called because it burns with no flame. Peat may be

mentioned as another natural fuel. It is composed of vegetable matter. In some kinds it is so far decomposed, or mineralised, that the eye does not detect vegetable fibres; when thoroughly dried, such peat breaks like a good lignite, and forms an excellent fuel.

METAMORPHIC ROCKS.

23. Quartz-rock, or quartzite, is an altered quartzose sandstone or grit; it is generally a white or grayish-yellow rock, very hard and compact. The original gritty character of the rock is distinct, but the granules appear as if they had been fused so far as to become mutually adherent. When the altered sandstone has been composed of grains of quartz, felspar, or mica, set in a siliceous, felspathic, or argillaceous base, we get a rock called greywacke, which is usually gray or grayish blue in colour.

24. Clay-slate is a grayish blue, or green, fine-grained hard rock, which splits into numerous more or less thin laminae, which may or may not coincide with the original bedding. Most usually the 'cleavage,' as this fissile structure is termed, crosses the bedding at all angles.

25. Crystalline limestone is an altered condition of common limestone. Saccharoid marble is one of the fine varieties: it frequently contains flakes of mica. Dolomite, or magnesian limestone, already described, is probably in many cases an altered limestone; the carbonate of lime having been partially dissolved out and replaced by carbonate of magnesia. Serpentine is also believed by some geologists to be a highly metamorphosed magnesian limestone.

26. Schists.--Under this term comes a great variety of crystalline rocks which all agree in having a foliated texture--that is to say, the constituent minerals are arranged in layers which usually, but not invariably, coincide with the original bedding. Amongst the schists come mica-schist (quartz and mica in alternate layers); chlorite-schist (chlorite with a little quartz, and sometimes with felspar or mica); talc-schist (talc with quartz or felspar); hornblende schist

(hornblende with a variable quantity of felspar, and sometimes a little quartz); gneiss (quartz, felspar, and mica).

27. General Character of Metamorphic Rocks.--All these rocks betray their aqueous origin by the presence of more or less distinct lines of bedding. They consist of various kinds of arenaceous and argillaceous deposits, which, under the influence of certain metamorphic actions, to be described in the sequel, have lost their original granular texture, and become more or less distinctly crystallised. And not only so, but their chemical ingredients have in many cases entered into new relations, so as to give rise to minerals which existed either sparingly or not at all in the original rocks. Frequently, it is quite impossible to say what was the original condition of some metamorphic rocks; often, however, this is sufficiently obvious. Thus, highly micaceous sandstones, as they are traced into a metamorphic region, are seen to pass gradually into mica-schist. When the bedding of gneiss becomes entirely obliterated, it is often difficult to distinguish that rock from granite, and in many cases it appears to pass into a true granite.

28. Granite is a crystalline compound of quartz, felspar (usually potash-felspar), and mica. Some geologists consider it to be invariably an igneous rock; but, as just stated, it sometimes passes into gneiss in such a way as to lead us to infer its metamorphic origin. There are certain areas of sandstone in the south of Scotland which are partially metamorphosed, and in these we may trace a gradual passage from highly baked felspathic sandstones with a sub-crystalline texture into a more crystalline rock which in places graduates into true granite. Granite, however, also occurs as an igneous rock.

29. Syenite is a crystalline compound of a potash-felspar and hornblende, and quartz is frequently present. Diorite is a crystalline aggregate of a soda-felspar and hornblende. Both syenite and diorite also occur as igneous rocks.

There are a number of other metamorphic rocks, but those mentioned are the most commonly occurring species.

IGNEOUS ROCKS.

30. Subdivisions.--In their chemical and mineralogical composition, igneous rocks offer great variety; but they all agree in having felspar for their base. They may be roughly divided into two classes, distinguished by the relative quantity of silica which they contain. Those in which the silica ranges from about 50 to 70 or 80 per cent. form what is termed the acidic group; while those in which the percentage of silica is less constitute the basic group of igneous rocks, so called because they contain a large proportion of the heavier bases, such as magnesia, lime, oxides of iron and manganese, &c. Igneous rocks vary in texture from homogeneous, compact, and finely crystalline masses up to coarsely crystalline aggregates, in which the crystals may be more than an inch in diameter. Sometimes they are dull and earthy in texture, at other times vesicular. When the vesicles are filled up with some mineral, the rock is said to be amygdaloidal, from the almond shape assumed by the kernels filling the cavities. When single crystals of any mineral are scattered through a rock, so as to be readily distinguished from the compact or crystalline base, the rock becomes porphyritic.

ACIDIC OR FELSPATHIC GROUP.

31. Trachyte (trachys, rough) is a pale or dark-gray rock, harsh and rough to the touch, in which felspar is the predominant mineral. It is a common product of eruption in modern volcanoes.

Clinkstone or phonolite is a greenish-gray, compact, felspathic rock, somewhat slaty or schistose, and weathers with a white crust. It gives a clear metallic sound under the hammer. It is a rock not met with among the older formations of the earth's crust, being confined to Tertiary (see table, p. 85) or still more recent times.

Obsidian or volcanic glass is usually black, brown, or green, and usually resembles a coarse bottle-glass. When it becomes vesicular, it passes gradually into the highly porous rock called pumice. It is eminently a geologically modern volcanic rock.

Felstone is a reddish-gray, bluish, greenish, or yellowish, hard, compact, flinty-looking rock, composed of potash-felspar and silica. It is generally splintery under the hammer. Some varieties are slaty, and are frequently mistaken for clinkstone, which they closely resemble. When the quartz in felstone is distinctly visible either as grains or crystals, the rock passes into a quartz-porphyry.

Granite is recognised as an igneous as well as a metamorphic rock. Sometimes the veins and dykes which proceed from or occur near a mass of granite contain no mica--this kind of rock is called elvan or elvanite.

Porphyrite or felspathite includes a number of rocks which have a felspathic base, through which felspar crystals are scattered more or less abundantly. Sometimes hornblende, or augite, or mica is present. The colour is usually dark--some shade of blue, green, red, puce, purple, or brown--and the texture varies from compact and finely crystalline up to coarsely crystalline. Porphyrites are usually porphyritic, and frequently amygdaloidal.

AUGITIC AND HORNBLENDIC OR BASIC GROUP.

32. Basalt is a dark or almost black compact homogeneous rock, composed of felspar and augite with magnetic iron. An olive-green mineral called olivine is very frequently present. The coarser-grained basalts are called dolerite. The columnar structure is not peculiarly characteristic of basalt. Many basalts are not columnar, and not a few columnar rocks are not basalts.

Greenstone or diorite is usually a dull greenish rock, sometimes gray, however, speckled with green. It is composed of soda-felspar and hornblende. The fine-grained compact greenstones are called aphanite.

Syenite, like granite, is recognised as an igneous as well as a metamorphic rock. There are several other rocks which come into the basic group, but those mentioned are the more common and typical species.

33. Fragmental Igneous Rocks.--All the igneous rocks briefly described above are more or less distinctly crystalline in texture. There is a class of igneous rocks, however, which do not present this character, but when fine-grained are dull and earthy in texture, and frequently consist merely of a rude agglomeration of rough angular fragments of various rocks. These form the FRAGMENTAL group of igneous rocks. The ejectamenta of loose materials which are thrown out during a volcanic eruption, consist in chief measure of fragments of lava, &c. of all sizes, from mere dust, sand, and grit, up to blocks of more than a ton in weight. These materials, as we shall afterwards see, are scattered round the orifice of eruption in more or less irregular beds. The terms applied to the varieties of ejectamenta found among modern volcanic accumulations, will be given and explained when we come to consider the nature of geological agencies. In the British Islands, and many other non-volcanic regions, we find besides crystalline igneous rocks, abundant traces of loose ejectamenta, which clearly prove the former presence of volcanoes. These materials are sometimes quite amorphous--that is to say, they shew no trace of water action--they have not been spread out in layers, but consist of rude tumultuous accumulations of angular and subangular fragments of igneous rocks. Such masses are termed trappean agglomerate and trappean breccia. At other times, however, the ejectamenta give evidence of having been arranged by the action of water, the materials having been sifted and spread out in more or less regular layers. What were formerly rude breccias and agglomerates of angular stones now become trappean conglomerates--the stones having been rounded and water-worn--while the fine ingredients, the grit, and sand, and mud, form the rock called trap tuff. Fragmental rocks are often quite indurated--the matrix being as hard as the included stones. But as a rule they are less hard than crystalline igneous rocks, and in many cases are loose and crumbling. When a fragmental rock is composed chiefly of rocks belonging to the acidic group, we say it is felspathic. When augitic and hornblendic materials predominate, then other terms are used; as, for example, dolerite tuff, greenstone tuff.

STRUCTURE AND ARRANGEMENT OF ROCK-MASSES.

34. The student can hardly learn much about the mineralogical composition of rocks, without at the same time acquiring some knowledge of the manner of their occurrence in nature. We have already briefly described certain sedimentary rocks, such as conglomerate, sandstone, and shale, and have in some measure touched upon their structure as rock-masses. These rocks, as we have seen, are arranged in more or less thick layers or beds, which are piled one on the top of the other. Rocks which are so arranged are said to be stratified, and are termed strata. We may also use the word stratum as an occasional substitute for bed. The planes of bedding or stratification are sometimes very close together, in other cases they are wide apart. When the separate beds are very thin, as in the case of shale, it is most usual to term them laminae, and to speak of the lamination of a shale, as distinguished from the bedding of a sandstone. Planes of bedding are generally more strongly marked than planes of lamination. The laminae frequently cohere, while beds seldom do. In the above figure, which represents a vertical cutting or section through horizontal strata, the planes of lamination are shewn at l, l, l, and those of stratification at s, s, s. There are hardly any limits to the thickness of a bed-- it may range from an inch up to many feet or yards, while laminae vary in thickness from an inch downwards.

35. Hitherto we have been considering the laminae and strata as lying in an approximately horizontal plane. Sometimes, however, the layers of deposition in a single stratum are inclined at various angles to themselves, as in the following figure. This structure is called false bedding; the layers or laminae not coinciding with the planes of stratification. It owes its origin to shifting currents, such as the ebb and flow of the tide, and very often characterises deposits which have been formed in shallow water. (Hillocks of drifting sand frequently shew a similar structure, but their false bedding is, as a rule, much more pronounced.)

36. Mud-cracks and Rain-prints.--The surfaces of some beds occasionally exhibit markings closely resembling those seen upon a flat sandy beach after

the retreat of the tide--hence they are called ripple-marks or current-marks. They are, of course, due to the gentle current action which pushes along the grains of sand, and hence, such marks may be formed wherever a current sweeps over the bottom of the sea with energy just sufficient for the purpose. But since the necessary conditions for the formation of ripple-mark occur most abundantly in shallow water, its frequent appearance in a series of strata may often be taken as evidence, so far, for the shallow-water origin of the beds. Besides ripple-marks we may also detect occasionally on the surfaces of certain strata mud-cracks and rain-prints. These occur most commonly in fine-grained beds, as in flagstones, argillaceous sandstones, shales, &c. The mud-cracks resemble those upon a mud-flat which are caused by the desiccation and consequent shrinkage of the mud when exposed to the sun. The old cracks have been subsequently filled up again by a deposition of mud or sand, usually of harder consistency than the rock traversed by the cracks. Hence, when the bed that overlies the mud-cracks is removed, we find a cast of these projecting from its under surface, or frequently the cast remains in its mould, and forms a series of curious ridges ramifying over the whole surface of the old mud-flat. Rain-prints are the small pits caused by the impact of large drops. They are usually deeper at one side than the other, from which we can infer the direction of the wind at the time the rain-drops fell. Like the mud-cracks, they are most commonly met with in fine-grained beds, and have been preserved in a similar manner. Some geologists have also been able to detect wave-marks, 'faint outlinings of curved form on a sandstone layer, like the outline left by a wave along the limit where it dies out upon a beach.'

37. Succession of Strata.--The succession of strata is often very diversified. Thus, we may observe in one and the same section numberless alternating beds of sandstone and shale from an inch or so up to several feet each in thickness, with seams of coal, fireclay, ironstone, and limestone interstratified among them. In other cases, again, the succession is simpler, and some deep quarries shew only one bed, as is the case with certain limestones, fine-grained sandstones (liver-rock), and many volcanic rocks. Some limestones, indeed, shew small trace of bedding throughout a vertical thickness of hundreds of feet.

38. Beds, their Extent, &c.--Beds of rock are not only of very different thicknesses, but they are also of very variable extent. Some may thin gradually away, or 'die out' suddenly, in a few feet or yards, while others may extend over many square miles. Beds of limestone, for example, can often be traced for leagues in several directions; and if this be the case with certain single beds, it is still more true of groups of strata. Thus the coal-bearing strata belonging to what is called the Carboniferous period cover large areas in Wales, England, Scotland, and Ireland, not less, probably, than 6000 square miles; and strata belonging to the same great period spread over considerable tracts on the Continent, and a very extensive area in North America. It holds generally true that beds of fine-grained materials are not only of more equal thickness throughout, but have also a wider extension than coarser-grained rocks. Fine sandstones, for example, extend over a wider area, and preserve a more equable thickness throughout than conglomerates, while limestones and coals are more continuous than either.

39. When a bed is followed for any distance it is frequently found to thin away, and give place to another occupying the same plane or horizon. Thus a shale will be replaced by a sandstone, a sandstone by a conglomerate, and vice versa. Sometimes also we may find a shale, as we trace it in some particular direction, gradually becoming charged with calcareous matter, so as by and by to pass, as it were, into limestone. Every bed must, of course, end somewhere, either by thus gradually passing into another, or by thinning out so as to allow beds which immediately overlie and underlie it to come together. Not unfrequently, however, a bed will stop abruptly, as in fig. 3.

40. Sequence of Beds.--It requires little reflection to see that the division plane between two beds may represent a very long period of time. Let the following diagram represent a section of strata, s being beds of grit, and a, b, c, beds of sandstone and shale. It is evident that the beds s must have been formed before the strata b were deposited above them. At x, the beds a and b come together, and were attention to be confined to that part of the section, the observer might be led to infer that no great space of time elapsed between the deposition of these two beds. Yet we see that an interval sufficient to allow of

the formation of the beds s must really have intervened. It is now well known that in many cases planes of bedding represent 'breaks in the succession' of strata--'breaks' which are often the equivalents of considerable thicknesses of strata. In one place, for example, we may have an apparently complete sequence of beds, as a, b, c, which a more extended knowledge of the same beds, as these are developed in some other locality, enables us to supplement, as a, s, b, c.

41. Joints.--Besides planes or lines of bedding, there are certain other division planes or joints by which rocks are intersected. The former, as we have seen, are congenital; the latter are subsequent. Joints cut right across the bedding, and are often variously combined, one set of joint planes traversing the rock in one direction, and another set or sets intersecting these at various angles. Thus, in many cases the rocks are so divided as easily to separate into more or less irregular fragments of various sizes. Besides these confused joints there are usually other more regular division planes, which intersect the strata in some definite directions, and run parallel to each other, often over a wide area: these are called master-joints. Two sets of master-joints may intersect the same strata, and when such is the case, the rock may be quarried in cuboidal blocks, the size of which will vary, of course, according as the two sets of joints are near or wide apart. Joints may either gape or be quite close; so close, indeed, as in many cases to be invisible to the naked eye. Certain igneous rocks frequently shew division planes which meet each other in such a way as to form a series of polygonal prisms. The basalt of Staffa and Giants' Causeway are familiar examples of this structure. Jointing is due to the gradual consolidation of the strata, and hence, in a series of strata, we may find the separate beds, according to their composition, very variously affected, some being much more abundantly jointed than others. Master-joints which traverse a wide district in some definite direction probably owe their origin to tension, the strata having been subjected to some strain by the underground forces.

42. Cleavage.--Fine-grained rocks, more especially those which are argillaceous, occasionally shew another kind of structure, which is called cleavage. Common clay-slate is a type of the structure. This rock splits up into

innumerable thin laminae or plates, the surface of which may either be somewhat rough, or as smooth nearly as glass. The cleavage planes, however, need not be parallel with the planes of bedding; in most cases, indeed, they cut right across these, and continue parallel to each other often over a very wide region. The original bedding is sometimes entirely obliterated, and in most cases of well-defined cleavage is always more or less obscure.

In the preceding diagram, the general phenomena of bedding, jointing, and cleavage are represented. The lines of bedding are shewn at S, S; another set of division-planes (joints) is observed at J, J, intersecting the former at right angles--A, B, C being the exposed faces of joints. The lines of cleavage are seen at D, D, cutting across the planes of bedding and jointing.

43. Foliation is another kind of superinduced structure. In a foliated rock the mineral ingredients have been crystallised and arranged in layers along either the planes of original bedding or those of cleavage. Mica-schist and gneiss are typical examples.

44. Concretions.--In many rocks a concretionary structure may be observed. Some sandstones and shales appear as if made up of spheroidal masses, the mineral composition of the spheroids not differing apparently from that of the unchanged rock. So in some kinds of limestone, as in dolomite, the concretionary structure is often highly developed, the rock resembling now irregular heaps of turnips with finger-and-toe disease, again, piles of cannon-balls, or bunches of grapes, and agglomerations of musket-shot. A spheroidal structure is occasionally met with amongst some igneous rocks. This is well seen in the case of rocks having the basaltic structure, in which the pillars, being jointed transversely, decompose along their division planes, so as to form irregular globular masses. In many cases, certain mineral matter which was originally diffused through a rock has segregated so as to form nodules and irregular layers. Examples of this are chert nodules in limestone; flint nodules in chalk; clay-ironstone balls in shale, &c.

45. Inclination of Strata.--Beds of aqueous strata must have been deposited in

horizontal or approximately horizontal planes; but we now find them most frequently inclined at various angles to the horizon, and often even standing on end. They sometimes, however, retain a horizontal position over a large tract of country. The angle which the inclined strata make with the horizon is called the dip, the degree of inclination being the amount of the dip; and a line drawn at right angles to the dip is called the strike of the beds. Thus, a bed dipping south-west will have a north-west and south-east strike. The crop or outcrop (sometimes also, but rarely, called the basset edge) of a bed is the place where the edge of the stratum comes to view at the surface. We may look upon inclined beds as being merely parts of more or less extensive undulations of strata, the tops of the undulations having been removed so as to expose the truncated edges of the beds. In the following diagram, for example, the outcrops of limestone seen at l, l, are evidently portions of one and the same stratum, the dotted lines indicating its former extent. The trough-shaped arrangement of the beds at s is called a synclinal curve, or simply a syncline; the arched strata at a forming, on the contrary, an anticlinal curve or anticline.

When strata shew many and rapid curves, they are said to be contorted. The diagram section (fig. 10) will best explain what is meant by this kind of structure.

46. In certain regions, the strata often dip in one and the same direction for many miles, at an angle approaching verticality, as in the following section. It might be inferred, therefore, that from A to B we had a gradually ascending series--that as we paced over the outcrop we were stepping constantly from a lower to a higher geological horizon. But, in such cases, the dip is deceptive, the same beds being repeated again and again in a series of great foldings of the strata. Such is the case over wide areas in the upland districts of the south of Scotland. The section (fig. 11) shews that the beds are actually inverted, the strata at x x being bent back upon strata which really overlie them.

47. Contemporaneous Erosion.--Occasionally a group of strata gives proof that pauses in the deposition of sediment took place, during which running water scooped out of the sediment channels of greater or less width, which

subsequently became filled up with similar or dissimilar materials. The diagram (fig. 12) will render this plain. At a we have beds of sandstone, which it is evident were at one time throughout as thick as they still are at x x. Having been worn away to the extent indicated, a deposition of clay (b) succeeded; and this, in turn, became eroded at c, c, the hollows being filled up again with coarse sand and gravel. In former paragraphs, we found reason to believe that lines of bedding indicated certain pauses in the deposition of strata. Here, in the present case, we have more ample proof in the same direction.

48. Unconformability.--But the most striking evidence of such pauses in the deposition of strata is afforded by the phenomenon called unconformability. When one set of rocks is found resting on the upturned edges of a lower set, the former are said to be unconformable to the latter. In the above section (fig. 13), a, a, are beds of sandstone resting on the upturned edges of beds of limestone, shale, and sandstone, l, s. Figs. 14 and 15 give other examples of the same appearance. It is evident that, in the case of fig. 14, the discordant bedding chronicles the lapse of a very long period. We have to conceive first of the deposition of the underlying strata in horizontal or approximately horizontal layers; then we have to think of the time when they were crumpled up into great convolutions, and the tops of the convolutions (the anticlines) were planed away: all these changes intervened, of course, after the lower set was deposited, and before the upper series was laid down. In the case represented in fig. 15, we have a double unconformability, implying a still more elaborate series of changes, and probably, therefore, a still longer lapse of time.

49. Overlap.--When the upper beds of a conformable group of strata spread over a wider area than the lower members of the same series, they are said to overlap. The accompanying diagram shews this appearance. An overlap proves that a gradual submergence of the land was going on at the time the strata were being accumulated. As the land disappeared below the water, the sediment gradually spread over a wider area, the more recently deposited sediment being laid down in places which existed as dry land at the time when the earliest accumulations were formed. Thus, in the accompanying illustration

(fig. 16), the stratum marked 1, resting unconformably upon older strata, is overlapped by 2, as that is by 3, and so on--all the beds in succession coming to repose upon the older strata at higher and higher levels, as the old land subsided.

50. Dislocations or Faults.--When strata, once continuous, have been broken across, and displaced or shifted along the line of breakage, they are said to be faulted, the fissure along which the displacement occurs being termed a fault or dislocation. The simplest form of a fault is that shewn in the following diagram, where strata of sandstone and shale, with a coal-seam, S, have been shifted along the line f. The direction in which the fault is inclined[D] is its hade, and the degree of vertical displacement of the beds is the amount of the dislocation. Generally, the beds seem to be pulled down in the direction of the downthrow, and drawn up on the opposite side of the fault, as shewn in the diagram. Sometimes the rocks on each side of a fault are smoothed and polished, and covered with long scratches, as if the two sides of the fissure had been rubbed together. This is the appearance called slickensides. Slickensides, however, may occur on the walls of a fissure which is not a displacement, but a mere joint or crack. A dislocation is spoken of as a downthrow or an upcast, according to the direction in which it is approached. Thus, a miner working along the coal-seam S, from a to b, would describe the fault, f, as an upcast, since he would have to mine to a higher level to catch his coal again. But, had he approached the fault from c to d, he would then have termed it a downthrow, because he would see from the hade of the fault that his coal-seam must be sought for at a lower level. Faults are of all sizes, from a foot or two up to vertical displacements of thousands of feet. Powerful dislocations can often be followed for many miles across a country, running in a more or less linear direction. Thus, one large fault has been traced across the breadth of Scotland, from near St Abb's Head, in the east, to the coast of Wigtown, in the west. Every large throw is accompanied by a number of smaller ones--some of which run parallel to the main fault, while many others seem to run out from this at various angles. Faults are of all geological ages. Some date back to a most remote antiquity, others are of quite recent origin; and no doubt faults are occurring even now. In the following diagram, the strata, a, a, have been

faulted and planed away before the strata, b, were deposited. Hence, in this case, it is evident that if we know the geological age of the beds, a and b, we can have an approximation to the age of the fault. If the beds, a, be Carboniferous, and those at b Permian, then we should say the fault was post-Carboniferous or pre-Permian.

[D] The degree of inclination is very variable. It may occur at almost any angle up to vertical. But, as a rule, the hade of the more powerful faults is steeper than that of minor displacements.

51. Metamorphic and Igneous Rocks--mode of their occurrence.--In the foregoing remarks on the structure and arrangement of rocks we have had reference chiefly to the aqueous strata--that is to say, the mechanically, chemically, and organically formed rocks. We were necessarily compelled, however, to make some reference to, and to give some description of, certain structures and arrangements which are not peculiar to aqueous strata, but characterise many metamorphic and igneous rocks as well. To avoid repetition it was also necessary, while treating of joints, &c., to give some account of certain structures which are the result of metamorphic action. But, for sake of clearness, we have reserved special account of the structure and mode of occurrence of metamorphic and igneous rocks to this place. After what has been said as to the structure and arrangement of aqueous strata, it is hardly needful to say much about the crystalline schists. These the student will understand to be merely highly altered aqueous rocks,[E] in which the marks of their origin are still more or less distinctly traceable. As a rule, metamorphic strata are contorted, twisted, and crumpled, although here and there comparatively horizontal stretches of altered rocks may be observed. The regions in which they occur are often hilly and mountainous, but this is by no means invariably the case. The greater part of the mountainous regions of the British Islands is occupied by rocks which are more or less altered; the more crystalline rocks, such as mica-schist, gneiss, &c., being abundantly developed in the Scottish Highlands, and in the north and west of Ireland; while those which are less altered cover large areas in the south of Scotland, and in Wales and the north-west of England. Throughout these wide areas the rocks

generally dip at high angles, and contortion and crumpling are of common occurrence. The finer-grained clay-rocks also exhibit fine cleavage planes, and are in some places quarried for roofing-slates--the Welsh quarries being the most famous. Here and there, bedding is entirely effaced, and the resulting rock is quite amorphous, and, becoming gradually more and more crystalline, passes at last into a rock which in many cases is true granite. The original strata have disappeared, and granite occupies their place, in such a way as to lead to the inference that the granite is merely the aqueous strata which have been fused up, as it were, in situ. At other times the granite would appear to have been erupted amongst the aqueous strata, for these are highly confused, and baked, as it were, at their junction with the granite, from which, also, long veins are seen protruding into the surrounding beds. Metamorphic granite, then, graduates, as a rule, almost imperceptibly into rocks which are clearly of aqueous origin; while on the contrary the junction-line between igneous granite and the surrounding rocks is always well marked. The origin of granite, however, is a difficult question, and one which has given rise to much discussion. Some further remarks upon the subject will be found in the sequel under the heading of Metamorphism.

[E] Igneous rocks have also in some cases undergone considerable alteration; fine-grained tuffs, for example, occasionally assume a crystalline texture.

52. True igneous rocks occur either in beds or as irregular amorphous masses. When they occur as beds interstratified with aqueous strata, they are said to be contemporaneous, because they have evidently been erupted at the time the series of strata among which they appear was being amassed. When, on the other hand, they cut across the bedding, they are said to be subsequent or intrusive, because in this case they have been formed at a period subsequent to the strata among which they have been intruded. The bed upon which a contemporaneous igneous rock reclines, often affords marks of having been subjected to the action of heat; sandstones being hardened, and frequently much jointed and cracked, owing to the shrinking induced by the heat of the once molten rock above, and clay-rocks often assuming a baked appearance. There is generally, also, some discoloration both in the pavement of rock upon

which the igneous mass lies, and in the under portions of the latter itself. The beds overlying a contemporaneous igneous rock, however, do not exhibit any marks of the action of heat; the old lava-stream having cooled before the sediment, now forming the overlying strata, was accumulated over its surface. One may often notice how the sand and mud have quietly settled down into the irregular hollows and crevices of the old lava, as in the following section, where i represents the igneous rock; a being the baked pavement of sandstone, &c.; and b the overlying sedimentary deposits. When the igneous rock itself is examined, its upper portions are often observed to be scoriaceous or cinder-like, and the under portions likewise frequently exhibit a similar appearance. It is generally most solid towards the centre of the bed. The vesicles, or pores, in the upper and lower portions are often flattened, and are frequently filled with mineral matter. Sometimes these cavities may have been filled at the time the rock was being erupted, but in most cases the mineral matter would appear to have been introduced subsequently by the action of water percolating through the rock. Occasionally we meet with igneous rocks which are more or less vesicular and amygdaloidal throughout their entire mass. Others, again, often shew no vesicular structure, but are homogeneous from top to bottom. The texture is also very variable, and this even in the same rock-mass; some portions being compact or fine-grained, and others coarsely crystalline. As a rule the rock is most crystalline towards the centre, and gets finer-grained as the top and bottom of the bed are approached. Not unfrequently, however, an igneous rock will preserve the same texture throughout. The jointing is also highly irregular as a rule. But in many cases, especially when the rock is fine-grained, the jointing is very regular. The basaltic columns of the Giants' Causeway and the Isle of Staffa are well-known examples of such regularly jointed masses. Igneous rocks frequently decompose into a loose earthy mass (wacke), and this is most markedly the case with those belonging to the basic group.

53. Contemporaneous igneous rocks are frequently associated with more or less regular beds of breccia, conglomerate, ash, tuff, &c. These are evidently the loose volcanic ejectamenta which accompanied former eruptions of lava, and have been arranged by the action of water. Beds of such materials,

however, frequently occur without any accompanying lava-form rocks. Nor are they always arranged in bedded masses. They sometimes appear filling vertical pipes which seem to have been the funnels of old volcanoes. The following section exhibits the general appearance of one of these volcanic necks. They are very common in some parts of Scotland, as in Ayrshire, and are frequently ranged along the line of a fault in the strata. Fig. 21 shews such a neck of ejectamenta, made up of fragments of various kinds of rock, such as sandstone, shale, limestone, coal, &c., sometimes without any admixture of igneous rocks. The strata through which the pipe has been pierced usually dip in towards the latter, and at their junction with the coarse agglomerate often shew marks of the action of heat, coal-seams having sometimes been 'burned' useless for a number of yards away from the 'neck.'

54. Intrusive igneous rocks occur as sheets, dykes, and necks. The sheets frequently conform for long distances to the bedding of the strata among which they occur, and are thus liable to be mistaken for contemporaneous rocks. But when they are closely examined, it will be seen that they not only bake or alter the beds above and below them, but seldom keep precisely to one horizon or level--occasionally rising to a higher, or sinking to a lower position in the strata, as shewn in the following diagram-section. Dykes are wall-like masses of igneous strata which cut across the strata, generally at a high angle (see d, d, fig. 22). In the neighbourhood of a recent volcanic orifice, numerous dykes are seen ramifying in all directions. In the British Islands some dykes have been followed in a linear direction for very long distances. Sometimes these occupy the sites of large dislocations, at other times they have cut through the strata without displacing them. Occasionally they appear to have been the feeders of the great sheets of igneous rock which here and there occur in their vicinity. The phenomena presented by the necks of intrusive rock do not differ from those characteristic of agglomerate or tuff necks. The strata are bent down towards the central plug of igneous rock, and are generally more or less altered at the line of junction.

55. Intrusive rocks offer, as a rule, some contrasts in texture to contemporaneous masses. They are seldom amygdaloidal, but when they are

so it is generally towards the centre of the mass. The kernels are usually minute and more or less spherical.

The diagram (fig. 23) shews the general mode of occurrence of igneous rocks on the large scale. The stratified aqueous deposits are indicated at a, a. These are overlaid by a series of alternating beds of crystalline (c) and fragmental (t) igneous rocks. An irregular intrusive sheet at i cuts across the beds a, a. At p, another intrusive mass is seen rising in a pipe, as it were, and overflowing the beds a, a, so as to form a cap. A volcanic neck filled with angular stones intersects the strata at n, and two dykes, approaching the vertical, traverse the bedded rocks at d, d. It will be noticed that the contemporaneous igneous rocks form a series of escarpments rising one above the other.

The alteration effected by igneous rocks is generally greatest in the case of intrusive masses. This is well seen in some of our coal-fields, where the coal has frequently been destroyed over large areas by the proximity of masses of what was once melted rock. It is curious to notice how the intrusive sheets in a great series of strata have forced their way along the lines of least resistance. Thus, in the Scottish coal-fields, we find again and again that intrusive sheets have been squirted along the planes occupied by coal-seams, these having been more easily attacked than beds of sandstone or shale. The coal in such cases is either entirely 'eaten up,' as it were, or converted into a black soot. At other times, however, it is changed into a kind of coke, while other seams at a greater distance from the intrusive mass have been altered into a kind of 'blind coal' or anthracite.

These remarks on the mode of occurrence of igneous rocks are meant to refer chiefly to those masses which occur in regions where volcanic action has long been extinct, as, for instance, in the British Islands. In the sequel, some account will be given of the appearances presented by modern volcanoes and volcanic rocks.

[1] It has been usual to apply the term trap or trappean rock to all the old igneous rocks which could neither be classed with the granites and syenites,

nor yet with the recent lavas, &c., which are connected with a more or less well-marked volcanic vent. The term trap (Swedish trappa, a flight of steps) was suggested by the terraced or step-like appearance presented by hills which are built up of successive beds of igneous rock. But the passage from the granitic into the so-called trap rocks, and from these into the distinctly volcanic, is so very gradual, that it is impossible to say where the one class ends and the other begins. The term trap, therefore, has no scientific precision, although it is sometimes very convenient as a kind of broad generic term to include a large number of rocks.

MINERAL VEINS.

56. The cracks and crevices and joint planes which intersect all rocks in a greater or less degree, are not unfrequently filled with subsequently introduced mineral matter, forming what are termed veins. This introduced matter may either be harder or less durable than the rock itself; in the former case, the veins will project from the surface of the stone, where that has been subjected to the weathering action of the atmosphere; in the latter case, the veins, under like circumstances, are often partially emptied of their mineral matter. Not unfrequently, however, the more or less irregularly ramifying, non-metalliferous veins appear as if they had segregated from the body of the rock in which they occur, as in the case of the quartz veins in granite. Besides these irregular veins, the rocks of certain districts are traversed in one or more determinate directions by fissures, extending from the surface down to unknown depths. These great fissures are often in like manner filled with mineral matter. The minerals are usually arranged in bands or layers which run parallel to the walls of the vein. Quartz, fluor-spar, barytes, calcite, &c. are among the commonest vein-minerals, and with these are frequently associated ores of various metals. A vein may vary in width from less than an inch up to many yards, and the arrangement of its contents is also subject to much variation. Instead of parallel layers of spars and ores, frequently a confused mass of clay and broken rocks, which are often cemented together with sparry matter, chokes up the vein. The ore in a vein may occur in one or more ribs, which often vary in thickness from a mere line up to masses several yards in

width. Sometimes the rocks are dislocated along the line of fissure occupied by a great vein; at other times no dislocation can be observed. Mineral veins, however, do not necessarily occupy dislocation fissures. They often occur in cavities which have been formed by the erosive action of acidulated water, in the way described in pars. 59, 60, and 61. This is frequently the case in calcareous strata. Such veins usually coincide more or less with the bedding of the rocks, but in the case of thick limestones they not unfrequently cut across the bedding in a vertical or nearly vertical direction, forming what are termed pipe-veins.

DYNAMICAL GEOLOGY.

57. Having considered the composition, structure, and arrangement of the rock-masses which form the solid crust of our globe, we have next to inquire into the nature of those physical agencies by the action of which the rocks, as we now see them, have been produced. The work performed by the various forces employed in modifying the earth's crust is at one and the same time destructive and reconstructive. Rocks are being continually demolished, and out of their ruins new rocks are being built. In other words, matter is constantly entering into new relations--now existing as solid rock, or in solution in water, or carried as the lightest dust on the wings of the wind; now being swept down by rivers into the sea, or brought under the influence of subterranean heat--but always changing, sooner or later, slowly or rapidly, from one form to another. The great geological agents of change are these: 1. THE ATMOSPHERE; 2. WATER; 3. PLANTS AND ANIMALS; 4. SUBTERRANEAN FORCES. We shall consider these in succession.

THE ATMOSPHERE.

58. All rocks have a tendency to waste away under the influence of the atmosphere. This is termed weathering. Under the influence of the sun's heat, the external portions of a rock expand, and again contract when they cool at night. The effect of this alternate expansion and contraction is often strikingly manifest in tropical countries: some rocks being gradually disintegrated, and

crumbling into grit and sand; others becoming cracked, and either exfoliating or breaking up all over their surface into small angular fragments. Again, in countries subject to alternations of extreme heat and cold, similar weathering action takes place. The chemical action of the atmosphere is most observable in the case of calcareous rocks. The carbonic acid almost invariably present acts as a solvent, so that dew and rain, which otherwise would in many cases have but feeble disintegrating power, are enabled to eat into such rocks as chalk and limestone, calcareous sandstones, &c. The oxygen of the atmosphere also unites with certain minerals, such as the proto-salts of iron, and converts them into peroxides. It is this action which produces the red and yellow ferruginous discolorations in sandstone. Chemical changes also take place in the case of many igneous rocks, the result being that a weathered 'crust' forms wherever such rocks are exposed to the action of the atmosphere. Of course, the rate at which a rock weathers depends upon its mineralogical and chemical composition. Limestones weather much more rapidly than clay-rocks; and augitic igneous rocks, as a rule, disintegrate more readily than the more highly silicated species. The weathering action of the atmosphere is also greatly aided by frost, as we shall see presently. The result of all this weathering is the formation of soil--soil being only the fine-grained debris of the weathered rocks. The angular debris found at the base of all cliffs in temperate and arctic regions, and on every hill and mountain which is subjected to alternations of extreme heat and cold, is also the effect of weathering. But these and other effects of frost will be treated of under the head of Frozen Water. The hillocks and ridges of loose sand (sand dunes) found in many places along the sea-margin, and even in the interior of some continents, as in Africa and Asia, are due to the action of the wind, which drives the loose grains before it, and piles them up. Sometimes also the wind carries in suspension the finest dust, which may be transported for vast distances before it falls to the ground. Thus, fine dust shot into the air by the volcanoes of Iceland has been blown as far as the Shetland Islands; and in tropical countries the dust of the dried-up and parched beds of lakes and rivers is often swept away during hurricanes, and carried in thick clouds for leagues. Rain falling through this dust soaks it up, and comes down highly discoloured, brown and red. This is the so-called blood-rain. Minute microscopic animal and vegetable organisms are often commingled with this dust, and falling into streams, lakes, or the sea, may thus become eventually buried in sediments very far removed from the place that gave them birth.

WATER.

59. The geological action of water in modifying the crust of the earth is twofold--namely, chemical and mechanical.

Underground Water.--All the moisture which we see falling as rain or snow does not flow immediately away by brooks and rivers to the sea. Some portion of it soaks into the ground, and finds a passage for itself by cracks and fissures in the rocks below, from which it emerges at last as springs, either at the surface of the earth, or at the bottom of the sea. Such are the more obvious courses pursued by the water--it flows off either by sub-aerial or subterranean channels. But a not inconsiderable portion soaks into the solid rocks themselves, which are all more or less porous and pervious. Water thus slowly soaking often effects very considerable chemical changes. Sometimes the binding matter which held the separate particles of the rock together is dissolved out, and the rock is thus rendered soft and crumbling; at other times, the reverse takes place, and the water deposits, in the minute interstitial pores, some binding matter by which the partially or wholly incoherent grains are agglutinated into a solid mass. Thus what were originally hard and tough rocks become disintegrated to such a degree, that they crumble to powder soon after they are exposed to the air; while some again are converted into a clay, and may be dug readily with a spade. And, on the other hand, loose sand is glued into a hard building-stone. There are many other changes effected upon rocks by water, in virtue of the chemical agents which it holds in solution. Indeed, it may be said that there are very few, if any, rocks in which the chemical action of interstitial water has not formerly been, or is not at present being, carried on. Besides that which soaks through the rocks themselves, there is always a large proportion of underground water, which, as we have said above, finds a circuitous route for itself by joints, cracks, and crevices. After coursing for, it may be, miles underground, such water eventually emerges as springs, which contain in solution the various ingredients which the water has chemically extracted from the rocks. These ingredients are then deposited in proportion as the mineral water suffers from evaporation. Water impregnated with carbonate of lime, for example, deposits that compound as soon as evaporation has carried off a certain percentage of the water itself, and the carbonic acid gas which it held. This is the origin of the mineral called travertine or calcareous tufa, which is so commonly met with on the margins of springs, rivers, and waterfalls.

60. Stalactites and stalagmites have been formed in a similar way. Water slowly oozing from the roof of a limestone cavern partially evaporates there, and a thin pellicle of carbonate of lime is formed; while that portion of the water which falls to the ground, and is there evaporated, likewise gives rise to the formation of carbonate of lime. By such constant dropping and evaporating, long tongue-and icicle-like pendants (stalactites) grow downwards from the roof; while at the same time domes and bosses (stalagmites) grow upwards from the floor, so as sometimes to meet the former and give rise to continuous pillars and columns. The great solvent power of carbonated water is shewn first by the chemical analysis of springs, and, secondly, by the great wasting effects which the long-continued action of these has brought about. Thus, it has been estimated that the fifty springs near Carlsbad, which yield eight hundred thousand cubic feet of water in twenty-four hours, contain in solution as much lime as would go to form a mass of stone weighing two hundred thousand pounds. Warm, or, as they are termed, thermal springs, frequently carry away with them, out of the bowels of the earth, vast quantities of mineral matter in solution. The waters at Bath, for instance, are estimated to bring to the surface an annual amount of various salts, the mass of which is not less than 554 cubic yards. One of the springs of Loueche, France, however, carries out with it no less than 8,822,400 pounds of gypsum annually, which is equal to about 2122 cubic yards.

61. It is easy to conceive, therefore, that in the course of ages great alterations must be caused by springs. Caves and winding galleries, and irregular channels, will be worn out of the rocks which are thus being dissolved.

Especially will this be the case in countries where calcareous rocks abound. It is in such regions, accordingly, where we meet with the most striking examples of caves and underground river-channels. The largest cave at present known is the Mammoth Cave, in Kentucky. This remarkable hollow consists of numerous winding galleries and passages that cross and recross, and the united length of which is said to be 217 miles. In calcareous countries, rivers, after flowing for, it may be, miles at the surface, suddenly disappear into the ground, and flow often for long distances before they reappear in the light of day. In some regions, indeed, nearly all the drainage is subterranean. The surface of the ground, in calcareous countries, frequently shews circular depressions, caused by the falling in of the roofs of caverns. Sometimes, also, great masses of rock, often miles in extent, get loosened by the dissolving action of subterranean water, and crash downwards into the valleys. Such landslips, as they are called, are not, however, confined to calcareous regions. In 1806, a large section of the Rossberg, a mountain lying to the north of the Righi, consisting of conglomerate overlying beds of clay, rushed down into the plains of Goldau, overwhelming four villages and nearly a thousand inhabitants. The cause of this catastrophe was undoubtedly the softening into mud of the clay-beds on which the conglomerate rested, for the season which had just terminated when the slip took place had been very wet. The mass of material that slid down was estimated to contain upwards of fifty-four millions of cubic yards; it reached not less than two and a half miles in length, by some three hundred and fifty yards wide, and thirty-five yards thick.

62. Surface-water--Rain.--Having now learned something as to the modifications produced by underground water, we turn next to consider the action of surface-water, and the results arising from that action. Rain, when it falls to the ground, carries with it some carbonic acid gas which it has absorbed from the atmosphere. Armed with this solvent, it attacks certain rocks, more especially limestones and chalk, a certain proportion of which it licks up and delivers over to brooks and streams. Under its influence, also, the finer particles of the soil are ever slowly making their way from higher to lower levels. Rocks which are being gradually disintegrated by weathering have their finer grains and particles, thus loosened, carried away by rain. Nor

is this rain-action so inconsiderable as might be supposed. In the gentler hollows of an undulating country, we frequently find accumulations of clay, loam, and brick-earth, which often reach many feet in thickness, and which are undoubtedly the results of rain washing down the particles of soil, &c. from the adjacent slopes.

63. River-action.--The water of streams and rivers almost invariably contains in solution one or more chemical compounds, and in this respect does not differ from the water of springs. Of course, this mineral matter is derived in considerable measure from springs, but is also no doubt to a large extent taken up by the rivers themselves, as they wash the rocks and soils on their journey to the sea. The amount of mineral matter thus transported must be something enormous, as is shewn by the chemical analyses of river-water. Bischof calculated that the Rhine carries in solution as much carbonate of lime as would suffice for the yearly formation of three hundred and thirty-two thousand millions of oyster-shells of the usual size--a quantity equal to a cube five hundred and sixty feet in the side, or a square bed a foot thick, and upwards of two miles in the side. But the mechanical erosion effected by running water is what impresses us most with the importance of rivers as geological agencies. This erosive action is due to the gravel, sand, and mud carried along by the water. These ingredients act as files in the hand of a workman, and grind, polish, and reduce the rocks against which they are borne. The beds of some streams that flow over solid rock are often pitted with circular holes, at the bottom of which one invariably finds a few rounded stones. These stones, kept in constant motion by the water, are the means by which the pot-holes, as they are called, have been excavated. When pot-holes are numerous, they often unite so as to form curious smooth-sided trenches and gullies. The same filing action goes on all over the bed of the stream wherever the solid rock is exposed. And while the latter is being gradually reduced, the stones and grit which act as the files are themselves worn and reduced; so that stones diminish in size, and grit passes into fine sand and mud, as they move from higher to lower levels. No doubt the erosive action of running water appears to have but small effect in a short time, and we are apt, therefore, to underestimate its power. But when our observations extend, we

see it is quite otherwise, and that, so far from being unimportant, running water is really one of the most powerful of all the geological agencies that are employed in modifying the earth's crust. Even within a comparatively short time, it is able to effect very considerable changes. Thus, the river Simeto, in Sicily, having become dammed by a stream of lava flowing from Etna, succeeded, in two hundred and fifty years, in cutting through hard solid basalt a new channel for itself, which measured from twenty to fifty metres in depth, and from twelve to eighteen in breadth. When, also, we remember the fact, that no river is absolutely free from mineral matter held in suspension, but that, on the contrary, all running water is more or less discoloured with sediment, which is merely the material derived from the disintegration of rocks, it will appear to us difficult to overestimate the power of watery erosion. To the mineral matter held in suspension, we have to add the coarser detritus, gravel and sand, which is being gradually pushed along the beds of rivers, and which, in the case of the Mississippi, has been estimated to equal a mass of seven hundred and fifty million cubic feet, discharged annually into the Gulf of Mexico. By careful measurements, it has also been ascertained that the same river carries down annually into the sea a weight of mud held in suspension which reaches the vast sum of 812,500,000,000 pounds. The total annual amount of mineral matter, whether held in suspension or pushed along the bottom of this great river, has been estimated to equal a mass 268 feet in height, with an area of one square mile.

64. Alluvium.--The sediment carried along and deposited by a river is called alluvium. Sometimes this alluvium covers wide areas, forming broad flats on one or both sides of a river, and in such cases it is due to the action of the floodwaters of the stream. Every time the river overflows the low grounds through which it passes, a layer of sediment is laid down, which has the effect of gradually raising the level of the alluvial tract. By and by a time comes when the river, which has all the while been slowly deepening its channel, is unable to flood the flats, and thereupon it begins to cut into these, and to form new flats at a somewhat lower level. In this way we often observe a series of alluvial terraces, consisting of gravel, sand, and silt, rising one above another along a river valley. Such are the terraces of the Thames and other rivers in

England, and of the Tweed, Clyde, Tay, &c. in Scotland. The great plains through which the Rhine flows between Basel and Bingen, are also well-marked examples of alluvial accumulations. There are very few streams, indeed, which have not formed such deposits along some portion of their course.

65. When a river enters a lake, the motion of the water is of course checked, and hence the heavier detritus, such as gravel and coarse sand, moves more slowly forward, and at last comes to rest on the bed of the lake, at no great distance from the mouth of the river. Finer sand and mud are carried out for some distance further, but eventually they also cease to move, and sink to the bottom. When the lake is sufficiently large, it catches all or nearly all the matter brought down by the river, which, as it issues from the lower end of the lake, is bright and clear. A well-known example of this phenomenon is that of the Rhone, which enters the Lake of Geneva turbid and muddy, but rushes out quite clear at the lower end of the lake. Lakes, therefore, are all being slowly or more rapidly silted up, and this, of course, is most conspicuous at the points where they are entered by rivers. Thus, at the head of the Lake of Geneva, it is manifest that the wide flat through which the river flows before it pours into the lake, has been conquered by the Rhone from the latter. In the times of the Romans, the lake, as we know, extended for more than one mile and a half further up the valley.

66. Deltas.--When there are no lakes to intercept fluviatile sediment, this latter is borne down to the sea, where it is deposited in precisely the same way as in a lake: the heavier detritus comes to rest first, the finer sediment being swept out for some distance further. So that, in passing from the river-mouth outwards, we have at first gravel, which gradually gets finer and finer until it is replaced by sand, while this in turn is succeeded by mud and silt. There is this difference, however, between lacustrine and fluvio-marine deposits, that while the former accumulate in water which is comparatively still, the latter are often brought under the influence of waves and currents, and become shifted and sifted to such a degree that fine and coarse detritus are frequently commingled; and there is, therefore, not the same orderly succession of coarse and fine

materials which characterises lacustrine deposits. Often, indeed, the currents opposite the mouth of a river are so strong, that little or no sediment is permitted to gather there. Usually, however, we find that rivers have succeeded in reclaiming more or less wide tracts from the dominion of the waves, or at all events have cumbered the bed of the sea with banks and bars of detritus. The broad plains formed at the mouth of a river are called deltas, from their resemblance to the Greek letter [Delta]. The deltas of the Nile, Ganges, and Mississippi are among the most noted. The term delta, however, is not exclusively applied to fluvio-marine deposits; rivers also form deltas in fresh-water lakes. It is usual, however, to restrict the term to extensive alluvial plains which are intersected by many winding channels, due to the rapid bifurcation of the river, which begins to take place at the very head of the great flat--that is to say, at the point where the river originally entered the sea (or lake).

67. Frozen Water.--We have now seen what can be done by the mechanical action of running water. We have next to consider what modifications are effected by freezing and frozen water. Water, as every one knows, expands in the act of freezing, and in doing so exerts great force. Let the reader bear in mind what has been said as to the passage of water through the minute and often invisible pores of rocks, and to its presence in cracks and crevices after every shower of rain, and he will readily see how excessive must be the waste caused by the action of frost. The water, to as great a depth as the frost extends, passes into the solid state, and in doing so pushes the grains of the rocks asunder, or wedges out large masses. No sooner does thaw ensue than the water, becoming melted, allows the grains of the rock to fall asunder; the outer skin of the rock, as it were, is disintegrated, and crumbles away, while fragments and masses lose their balance in many cases, and topple down. Hence it is, that in all regions where frost acts, the hill-tops and slopes are covered with angular fragments and debris, and a soil is readily formed by the disintegration of the rocks.

River-ice is often a potent agent of geological change. Stones get frozen in along the margins of a river, and often debris falls down from cliff and scaur upon the surface of the ice; when thaw sets in, and the ice breaks up, stones and rubbish are frequently floated for long distances, and may even be carried out to sea before their support fails them, and they sink to the bottom. In some cases, when the ice is very thick, it may run aground in a river, and confuse and tumble up the deposits gathering at the bottom. Ice sometimes forms upon stones at the bottom of a river, and floats these off; and this curious action may take place even although no ice be forming at the time on the surface of the water.

68. Glaciers, Icebergs, and Ice-foot.--In certain mountainous districts, and in arctic and antarctic regions, snow accumulates to such an extent that its own weight suffices to press the lower portions into ice. Alternate thawing and freezing also aid in the formation of the ice, which soon begins to creep down the mountain-slopes into the valleys, where it constitutes what are called glaciers or ice-rivers. These great masses of ice attain often a great thickness, and frequently extend for many miles along the course of a valley. In the Alps they occasionally reach as much as five hundred or six hundred feet in depth. In Greenland, however, there are glaciers probably not less than five thousand feet thick; and the glacier ice of the antarctic continent has been estimated even to reach twelve miles in thickness. Glaciers flow slowly down their valleys, at a rate which varies with the slope of their beds and the mass of the ice. Some move only a few inches, others two or three feet, in a day. Their forward motion is arrested at a point where the ice is melted just as fast as it comes on. A glacier is always more or less seamed with yawning cracks, which are called crevasses. These owe their origin to the unequal rate at which the different parts of the ice flow; this differential motion causing strains, to which the ice yields by snapping asunder. The flanks of a glacier are usually fringed with heaps of angular blocks and debris which fall from the adjacent rocky slopes, and some of this rubbish tumbling into the gaping crevasses must occasionally reach to the bottom of the ice. The rubbish heaps (superficial moraines) travel slowly down the valley on the surface of the ice, and are eventually toppled over the end of the glacier, where they form great banks and mounds. These are called terminal moraines. The rocky bed of a glacier is invariably smoothed and polished, and streaked with coarse and fine

striae, or scratches, which run parallel to the direction of the ice-flow. These are due to the presence, at the bottom of the ice, of such angular fragments as become detached from the underlying rocks, or of boulders and rubbish which have been introduced from above. The stones are ground by the ice along the surface of its bed, causing ruts and scratches, while the finer material resulting from the grinding action forms a kind of polisher. The stones acting as gravers are themselves covered with striae, and their sharp edges get smoothed away. In alpine districts there is always a good deal of water circulating underneath a glacier, and this washes out the sand and fine clay. Thus it is that rivers issuing from glaciers are always more or less discoloured brown, yellow, green, gray, or blue, according to the nature of the rocks which the ice has pounded down into mud. In Greenland many of the large glaciers go right out to sea, and owing to their great thickness are able to dispossess the sea sometimes for miles. But erelong the greater specific gravity of the sea-water forces off large segments from the terminal front of the ice, which float away as icebergs. Large masses are also always falling down from the ice-front. Occasionally, big blocks and debris are floated away on the icebergs, but this does not appear to be common. In Greenland there is very little rock-surface exposed, from which blocks can be showered down upon the glaciers, and the surface of the latter is therefore generally free from superficial moraines. A kind of submarine terminal moraine, however, gathers in front of some glaciers, made up chiefly of the stones and rubbish that are dragged along underneath the ice, and exposed by the breaking-off of icebergs, but partly composed also of the sand and mud washed out by sub-glacial waters. A narrow belt of ice forms along the sea-coast in arctic regions, which often attains a thickness of thirty or forty feet. This is called the ice-foot. It becomes loaded with debris and blocks, which fall upon it from the cliffs above; and, as large portions are frequently detached from the cliffs in summer-time, they sail off with their cargoes of debris, and drop these over the sea-bottom as they gradually melt away. The ice-foot is the great distributor of erratics or wandered blocks, the part taken in this action by the huge icebergs which are discharged by the glaciers being, comparatively speaking, insignificant. But when these latter run aground, they must often cause great confusion among the beds of fine material accumulating upon the floor of the sea.

69. The Sea.--Sea-water owes its saltness to the presence of various more or less soluble substances, such as common salt, gypsum, Epsom salts, chloride of magnesium, &c. Besides these, there are other ingredients held in solution, which, although they can be detected in only minute quantities in sea-water, are yet of the very utmost importance to marine creatures. This is the case with carbonate of lime, vast quantities of which are carried down by many rivers to the sea. But it must be nearly all used up in the formation of hard shells and skeletons by molluscs, crustaceans, corals, &c., for very little can be traced in the water itself. Silica is also met with sparingly, and is abstracted by some creatures to form their hard coverings.

70. Breaker-action--Currents.--The most conspicuous action of the sea, as a geological agent, takes place along its margin, where the breakers are hurled against the land. Stones and gravel are borne with more or less intense force against the rocks, and by their constant battering succeed eventually in undermining the cliffs, which by and by become top-heavy, and large masses fall down and get broken up and pounded into gravel and sand. The new wall of rock thus exposed becomes in turn assaulted, and in course of time is undermined in like manner. The waste of the cliffs is greatly aided by the action of frost, which loosens the jointed rocks, and renders them an easier prey to the force of the waves. Of course, the rapidity with which a coast-line is eaten into depends very much upon the nature of the rocks. Where these are formed of loose materials like sand, gravel, or clay, considerable inroads are effected by the sea in a comparatively short time. Thus, along some parts of the English coast, as between Flamborough Head and the mouth of the Humber, and between the Wash and the Thames, it is estimated that the land is wasted away at the rate of a yard per annum. Where hard rocks form the coast-line the rate of waste is often exceedingly slow, and centuries may elapse without any apparent change being effected. When the rocks are of unequal hardness the coast-line becomes very irregular, the sea carving out bays and gullies in the softer portions, while the more durable masses stand out as capes and bold headlands. Not unfrequently, such headlands are converted into sea-stacks and rocky islets, as one may observe along the rockier parts of our

shore-lines. Close inshore, the bulkier debris derived from the waste of the land often accumulates, forming beds and banks of shingle and gravel. The finer materials are carried farther out to sea, and distributed over the sea-floor by the action of the tide and currents. Tidal and other currents may also have some denuding effect upon the sea-bottom, but this can only be in comparatively shallow water. The great bulk of the material derived from the waste of the coasts by the mechanical action of the breakers, travels for no great distance. But the fine mud brought down by rivers is frequently transported for vast distances before it settles. So fine, indeed, is some of this sedimentary material, that it may be carried in suspension by sea-currents for thousands of miles before it sinks to the bottom.

71. From this short outline it becomes evident, therefore, that the coarser-grained the deposit, the smaller will be the area it covers; while conversely, the finer the accumulation, the more widely will it be distributed. A partial exception to this rule is that of the debris scattered over the bottom of the ocean by icebergs and detached portions of ice-foot. These are often floated for vast distances by currents before they finally melt away, and hence the coarse debris transported by them must be very widely distributed over that part of the sea-bottom which is traversed by currents flowing out of the Arctic and Antarctic Oceans. Although the deeper recesses of the ocean appear to be covered only with ooze and fine mud, yet in some instances coarse sand, and even small stones, have been brought up from depths of a hundred fathoms, so that currents may occasionally carry coarser materials for great distances from the shore. The shifting action of tidal currents succeeds in giving rise to very irregular deposits in shallow seas. The soundings often shew sudden changes from gravel to sand and mud, nor can there be any doubt that, could we lay bare the sea-bottom, we should often observe gravel shading off into sand, and sand into mud, and vice versa. But as we receded from the shore, and approached areas which were once deeply submerged, we should find that the change of material was generally from coarse to fine.

GEOLOGICAL ACTION OF PLANTS AND ANIMALS.

72. Plants.--The disintegration of rocks is often aided by the action of plants, which force their roots into joints and crevices, and thus loosen blocks and fragments. Carbonic acid, derived from the decay of plants, being absorbed by rain-water, acts chemically upon many rocks, as in the case of limestone (see 59, 60, 61). In temperate regions, vegetation frequently accumulates, under certain conditions, to form very considerable masses. Of such a nature is peat, which, as is well known, covers many thousands of acres in the British Islands. This substance is composed fundamentally of the bog-moss (Sphagnum palustre), with which, however, are usually associated many other marsh-loving plants. The lower parts of bog-moss die and decay while its upper portions continue to flourish, and thus, in process of time, a thickness of peat is accumulated to the extent of six, twelve, twenty-four, or even forty feet. Many of the hill-tops and hill-slopes in Scotland and Ireland are covered with a few feet of peat, but it is only in valleys and hollows where the peat-bogs attain their greatest depth. In not a few cases, the bogs seem to occupy the sites of ancient lakes, shell-marl often occurring at the bottom of these. The trunks and roots of trees are also commonly met with underneath peat, and occasionally the remains of land animals. Frequently, indeed, it would seem as if the overthrow of the trees, by obstructing the drainage of the country, had given rise to a marsh, and the consequent formation of peat. Some of the most valuable peat closely resembles lignite, and makes a good fuel. In tropical countries, the rapidity with which vegetation decays prevents, as a rule, any great accumulation taking place; but the mangrove swamps are exceptions.

73. Animals.--The action of animal life is for the most part conservative and reconstructive. Considerable accumulations of shell-marl take place in fresh-water lakes, and the flat bottoms which mark the sites of lakes which have been drained are frequently dug to obtain this material. But by far the most conspicuous formations due to the action of animal life accumulate in the sea. Molluscs, crustaceans, corals, and the like, secrete from the ocean the carbonate of lime of which their hard shells and skeletons are composed, and these hard parts go to the formation of limestone. The most remarkable masses of modern limestone occur within intertropical regions. These are the coral reefs of the Pacific and Indian Oceans.

74. Coral is the calcareous skeleton of certain small soft-bodied gelatinous animals called actinozoa. These zoophytes flourish only in clear water, the temperature of which is not below 66 deg. F., and they cannot live at greater depths than one hundred feet. There are three kinds of coral reef--namely, fringing reefs, barrier reefs, and atolls. Fringing reefs occur, as a rule, near to the shore; but if this latter be gently sloping, they may extend for one or even two miles out to sea; as far, indeed, as the depth of water is not too great for the actinozoa. Barrier reefs are met with at greater distances from the land, and often rise from profound depths. The barrier reef which extends along the north-east coast of Australia, often at a distance from the land of fifty or sixty miles, stretches, with interruptions, for about 1250 miles, with a breadth varying from ten to ninety miles. In some places, the depth of the sea immediately outside of this reef exceeds 1800 feet. Sometimes barrier reefs completely encircle an island or islands, which are usually mountainous, as in the case of Pouynipete, an island in the Caroline Archipelago, and the Gambier Islands in the Low Archipelago. Atolls are more or less irregular ring-shaped reefs inclosing a lagoon of quiet water. They usually rise from profound depths; Keeling Atoll, in the Indian Ocean, is a good example. The upper surface of atolls and barrier reefs often peers at separate points above the level of the sea, so as to form low-lying islets. In some cases, the land thus formed is almost co-extensive with the reef, and being clothed with palms and tropical verdure, resembles a beautiful chaplet floating, as it were, in mid-ocean. The rock of a coral reef is a solid white limestone, similar in composition to that of the limestones occurring in this country. In some places, it is quite compact, shewing few or no inclosed shells or other animal remains; in other places, it is made up of broken and comminuted corals cemented together, or of masses of coral standing as they slowly grew, with the spaces between the separate clumps filled up with coral sand and triturated fragments and grit of coral and shell. The thickness of the reefs is often very great, reaching in many cases to thousands of feet. At the Fijis, the reef can hardly be less than 2000 or 3000 feet thick. Below a depth of one hundred feet, all the coral rock is dead, and since the coral zoophytes do not live at greater depths than this, it follows that the bed of the sea in which coral reefs occur must have slowly subsided during

a long course of ages. Mr Darwin was the first to give a reasonable explanation of the origin of coral reefs. Briefly stated, his explanation is as follows: The corals began to grow first in water not exceeding one hundred feet in depth, and built up to the surface of the sea, thus forming a fringing reef at no great distance from the land. This initial step is shewn at A, B, in the accompanying section across a coral island. A, A, are the outer edges of the fringing reef; B, B, the shores of the island; and S1 the level of the sea. Subsidence ensuing, the island and the sea-bottom sink slowly down, while the coral animals continue to grow to the surface--the building of the reef keeping pace with the subsidence. By and by the island sinks to the level S2, when B', B', represent the shores of the now diminished island, and A', A', the outer edges of the reef, which has become a barrier reef; C, C, being the lagoon between the reef and the central island. We have now only to suppose a continuance of the submergence to the level S3, when the island disappears, its site being occupied by a lagoon, C'--the reef, which has at the same time become an atoll, being shewn at A", A".

75. In extra-tropical latitudes, great accumulations of carbonate of lime are also taking place. The bottom of the Atlantic has been found to be covered, over vast areas, by a fine calcareous sticky deposit called ooze, which would appear to consist for the most part of the skeletons of minute animal organisms, called Foraminifera. This accumulation, when dried, closely resembled chalk, and there can be no doubt that in the deep recesses of the Atlantic we have thus a gradually increasing deposit of carbonate of lime, which rivals, if it does not exceed, in extent the most widely spread calcareous rocks with which we are acquainted. A small percentage of siliceous materials occurs in the ooze, made up partly of granules of quartz, and partly of the skeletons and coverings of minute animal and vegetable organisms. When in process of time the chemical forces begin to act upon the siliceous matter diffused through the Atlantic ooze, segregation, or the gathering together of the particles, may take place, and nodules of flint will be the result, similar to the flint nodules which occur in chalk, and the cherty concretions in limestones. Animalcules with siliceous envelopes and skeletons are by no means so abundant as those that secrete carbonate of lime, but they are very widely diffused through the oceans,

and in favourable places are so abundant that they may well give rise eventually to extensive beds of flint. Ehrenberg calculated that 17,946 cubic feet of these organisms were formed annually in the muddy bottom of the harbour at Wismar, in the Baltic.

It would appear from recent observations (Challenger expedition) that the calcareous ooze at the bottom of the Atlantic and Southern Oceans, which occurs at a mean depth of 2250 fathoms, passes gradually as the ocean deepens into a gray ooze, which is less calcareous, and which occurs at a mean depth of 2400 fathoms. At still greater depths this gray ooze also disappears, and is replaced by red clay at a mean depth of 2700 fathoms. The minute creatures (foraminifera and pelagic mollusca chiefly) whose shells go to form the calcareous ooze, live for the most part on the surface, and swarm all over the areas in which ooze and red clay occur at the bottom. Hence it seems probable that the clay is merely the insoluble residue or ash, as it were, of the organisms--the delicate shells, as they slowly sink to the more profound depths, being dissolved by the free carbonic acid, which, as observations would seem to shew, occurs rather in excess at great depths. Thus we see how the organic forces may give rise to extensive accumulations of inorganic matter, closely resembling the finest silt or mud which is carried down to the sea by rivers, and distributed far and wide by ocean currents.

SUBTERRANEAN FORCES.

76. There have been many speculations as to the condition of the interior of the earth. Some have inferred that the external crust of the globe incloses a fluid or molten mass; others think it more probable that the interior is solid, but contains scattered throughout its bulk, especially towards the surface of the earth, irregular seas of molten matter, occupying large vesicles or tunnels in the solid honey-combed mass. At present, the facts known would appear to be best explained by the latter hypothesis. All that we know from observation is, that the temperature increases as we descend from the surface. The rate of increase is very variable. Thus, in the Artesian well at Neuffen, in Wuertemberg, it was as much as 1 deg. F. for every 19 feet. In the mines of

Central Germany, however, the increase is only 1 deg. F. for every 76 feet; while in the Dukinfield coal-pit, near Manchester, the increase was still less, being only 1 deg. F. in 89 feet. Taking the average of many observations, it may be held as pretty well proved that the temperature of the earth's crust increases 1 deg. for every 50 or 60 feet of descent after the first hundred.

77. The crust of the earth is subject to certain movements, which are either sudden and paroxysmal, or protracted and tranquil. The former are known as earthquakes, which may or may not result in a permanent alteration of the relative level of land and sea; the latter always effect some permanent change, either of upheaval or depression.

78. Earthquakes have been variously accounted for. Those who uphold the hypothesis of a fluid interior think the undulatory motion experienced at the surface is caused by movements in the underlying molten mass--an earthquake being thus 'the reaction of the liquid nucleus against the outer crust.' By others, again, earthquakes are supposed to be caused by the fall of large rock-masses from the roofs of subterranean cavities, or by any sudden impulse or blow, such as might be produced by the cracking of rocks in a state of tension, by a sudden volcanic outburst, or sudden generation or condensation of steam. In support of this latter hypothesis, many facts may be adduced. The undulatory motion communicated to the ground during gunpowder explosions, or by the fall of rocks from a mountain, is often propagated to great distances from the scene of these catastrophes, and the phenomena closely resemble those which accompany a true earthquake. When the level of a district has been permanently affected by an earthquake, the movement has generally resulted in a lowering of the surface. Thus, in 1819, the Great Runn of Cutch, in Hindustan, was depressed over an area of several thousand square miles, so as during the monsoons to become a salt lagoon. Occasionally, however, we find that elevation of the land has taken place during an earthquake. This was the case in New Zealand in 1855, when the ground on which the town of Wellington stands rose about two feet, and a cape in the neighbourhood nearly ten feet. Sometimes the ground so elevated is, after a shorter or longer period, again depressed to its former level. A good example of this occurred in South

America in 1835. The shore at Concepcion was raised a yard and a half; and the Isle Santa Maria was pushed up two and a half yards at one end, and three and a half yards at the other. But only a few months afterwards the ground sank again, and everything returned to its old position. The heaving and undulatory motion of an earthquake produces frequently considerable changes at the surface of the ground, besides an alteration of level. Rocks are loosened, and sometimes hurled down from cliff and mountain-side, and streams are occasionally dammed with the soil and rubbish pitched into them. Sometimes also the ground opens, and swallows whatever chances to come in the way. If these chasms close again permanently, no change in the physiography of the land may take place, but sometimes they remain open, and affect the drainage of the country.

79. Movements of Upheaval and Depression.--Besides the permanent alteration of level which is sometimes the result of a great earthquake, it is now well known that the crust of the earth is subject to long-continued and tranquil movements of elevation and depression. The cause of these movements is at present merely matter for speculation, some being of opinion that they may be caused by the gradual contraction of the slowly cooling nucleus of the earth, which would necessarily give rise to depression, while this movement, again, would be accompanied by some degree of elevation-- the result of the lateral push or thrust effected by the descending rock-masses. It is doubtful, however, if this hypothesis will explain all the appearances. The Scandinavian peninsula affords a fine example of the movements in question. At the extremity of the peninsula (Scania), the land is slowly sinking, while to the north of that district gradual elevation is taking place at a very variable rate, which in some places reaches as much as two or three feet in a century. Movements of elevation are also affecting Spitzbergen, Northern Siberia, North Greenland, the whole western borders of South America, Japan, the Kurile Islands, Asia Minor, and many other districts in the Mediterranean area, besides various islets in the great Pacific Ocean. The proofs of a slow movement of elevation are found in old sea-beaches and sea-caves, which now stand above the level of the sea. In the case of Scandinavia, it has been noticed that the pine-woods which clothe the mountains are being slowly elevated to

ungenial heights, and are therefore gradually dying out along their upper limits. The proofs of depression of the land are seen in submerged forests and peat, which occur frequently around our own shores, and there is also strong human testimony to such downward movements of the surface. The case of Scania has already been referred to. Several streets in some of its coast towns have sunk below the sea, and it is calculated that the Scanian coast has lost to the extent of thirty-two yards in breadth within the past hundred and thirty years. The coral reefs of southern oceans also afford striking evidence of a great movement of depression.

Not long ago a theory was started by a French savant, M. Adhemar, to account for changes in the sea-level, without having recourse to subterranean agency. He pointed out that a vast ice-cap, covering the northern regions of our hemisphere, as was certainly the case during what is termed the glacial epoch, would cause a rise of the sea by displacing the earth's centre of gravity. Mr James Croll has recently strongly supported this opinion; and there can be no doubt that we have here a vera causa of considerable mutations of level. It is unquestionably true, however, that great oscillatory movements, such as described above, and which can only be attributed to subterranean agencies, have frequently taken and are still taking place.

80. Such movements of the earth's crust cannot take place without effecting some change upon the strata of which that crust is composed. During depression of the curved surface of the earth, the under strata must necessarily be subjected to intense lateral pressure, since they are compelled to occupy less space, and contortion and plication will be the result. It is evident also that contortion will diminish from below upwards, so that we can conceive that excessive contortion may be even now taking place at a great depth from the surface in Greenland. During a movement of elevation, on the other hand, the strata are subjected to excessive tension, and must be seamed with great rents: when the elevating force is removed, the disrupted rocks will settle down unequally--in other words, they will be faulted, and their continuity will be broken. But both contortion and faulting may be due, on a small scale, to local causes, such as the intrusion of igneous rocks, the consolidation of strata, the

falling in of old water-courses, &c. Cleavage is believed to have been caused by compression, such as the rocks might well be subjected to during great movements of the earth's crust. The particles of which the rock is composed are compressed in one direction, and of course are at the same time drawn out at right angles to the pressure. This is observed not only as regards the particles of the rock themselves, but imbedded fossils also are distorted and flattened in precisely the same way.

81. Volcanoes.--Besides movements of elevation and depression, there are certain other phenomena due to the action of the subterranean forces. Such are the ejection from the interior of the earth of heated matters, and their accumulation upon the surface. The erupted materials consist of molten matter (lava), stones and dust, gases and steam--the lava, ashes, and stones gradually accumulating round the focus of ejection, and thus tending to form a conical hill or mountain. Could we obtain a complete section of such a volcanic cone, we should find it built up of successive irregular beds of lava, and layers of stones and ashes, dipping outwards and away from the source of eruption, but having round the walls of the crater (that is, the cavity at the summit of the truncated cone) a more or less perceptible dip inwards. Fig. 25 gives a condensed view of the general phenomena accompanying an eruption. In this ideal section, a is the funnel or neck of the volcano filled with lava; b, b, the crater. The molten lava is highly charged with elastic fluids, which continually escape from its surface with violent explosions, and rise in globular clouds, d, d, to a certain height, after which they dilate into a dark cloud, c. From this cloud showers of rain, e, are frequently discharged. Large and small portions of the incandescent lava are shot upwards as the imprisoned vapour of water explodes and makes its escape, and, along with these, fragments of the rocks forming the walls of the crater and the funnel are also violently discharged; the cooled bombs, angular stones, and lapilli, as the smaller stones are called, falling in showers, f, upon the exterior parts of the cone or into the crater, from which they are again and again ejected. Most frequently the great weight of the lava inside the crater suffices to break down the side of the cone, and the molten rock escapes through the breach. Sometimes, however, it issues from beneath the base of the cone. At other times, finding for itself some weak place

in the cone, it may flow out by a lateral fissure, g. In the diagram, i, i represents the lava streaming down the outward slopes, jets of steam and fumaroles escaping from almost every part of its surface. Forked lightning often accompanies an eruption, and is supposed to be generated by the intense mutual friction in the air of the ejected stones. The trituration to which these are subjected reduces them, first, to a kind of coarse gravel (lapillo); then to sand (puzzolana); and lastly, to fine dust or ashes (ceneri).

82. Lava.--Any rock which has been erupted from a volcano in a molten state is called lava. Some modern lava-streams cover a great extent of surface. One of two streams which issued from the volcano of Skaptur Jokul (Iceland) in 1783 overflowed an area fifty miles in length, with a breadth in places of fifteen; the other was not much less extensive, being forty miles in length, with an occasional breadth of seven. In some places the lava exceeded 500 feet in thickness. Again, in 1855, an eruption in the island of Hawaii sent forth a stream of lava sixty-five miles long, and from one to ten miles wide. The surface of a stream quickly cools and consolidates, and in doing so shrinks, so as to become seamed with cracks, through which the incandescent matter underneath can be seen. As the current flows on, the upper crust separates into rough ragged scoriform blocks, which are rolled over each other and jammed into confused masses. The slags that cake upon the face or front of the stream roll down before it, and thus a kind of rude pavement is formed, upon which the lava advances and is eventually consolidated. Thus, in most cases, a bed of lava is scoriaceous as well below as above. Other kinds of lava are much more ductile and viscous, and coagulate superficially in glossy or wrinkled crusts. When lava has inclosed fragments of aqueous rocks, such as limestone, clay, or sandstone, these are observed to have undergone some alteration. The sandstone is often much hardened, the clay is porcelainised, and the limestone, still retaining its carbonic acid, assumes a crystalline texture. But the aqueous rock upon which lava has cooled does not usually exhibit much change, nor does the alteration, as a rule, extend more than a few feet (often only a few inches) into the rock. A lava-current which entered a lake or the sea, however, has sometimes caught up much of the sediment gathering there, and become so commingled with it, that in some parts it is hard to say whether the resulting

rock is more igneous or aqueous. Lava which has been squirted up from below into cracks and crevices, and there consolidated so as to form dykes, sometimes, but not often, produces considerable alteration upon the rocks which it intersects. The basaltic structure is believed to be due to the contraction of lava consequent upon its cooling. The axes of the prisms are always perpendicular to the cooling surface or surfaces, and in some cases the columns are wonderfully regular. There are numerous varieties of lava, such as basalt, obsidian, pitchstone, pearlstone, trachyte, &c.; some are heavy compact rocks, others are light and porous. Many are finely or coarsely crystalline; others have a glassy and resinous or waxy texture. Some shew a flaky or laminated structure; others are concretionary. Most of the lava rocks, however, are granularly crystalline. In many, a vesicular character is observed. These vesicles, being due to the bubbles of vapour that gathered in the molten rock, usually occur in greatest abundance towards the upper surface of a bed of lava. They are also more or less well developed near the bottom of a bed, which, as already explained, is frequently scoriaceous. Occasionally the vesicles are disseminated throughout the entire rock. As a rule, those lavas which are of inferior specific gravity are much more vesicular than the denser and heavier varieties. The vesicles are usually more or less flattened, having been drawn out in the direction in which the lava-current flowed. Sometimes they are filled, or partially filled, with mineral matter introduced at the time of eruption, or subsequently brought in a state of solution and deposited there by water filtering through the rock: this forms what is called amygdaloidal lava. In volcanic districts, the rocks are often traversed by more or less vertical dykes or veins of igneous matter. These dykes appear in some cases to have been formed by the filling up of crevices from above--the liquid lava having filtered downwards from an overflowing mass. In most cases, however, the lava has been injected from below, and not unfrequently the 'dykes' seem to have been the feeders from which lava-streams have been supplied--the feeders having now become exposed to the light of day either by some violent eruption which has torn the rocks asunder, or else by the gradual wearing away of the latter by atmospheric and aqueous agencies.

METAMORPHISM.

83. Mention has already been made of the fact, that the heated matters ejected from volcanoes, or forcibly intruded into cracks, crevices, &c., occasionally alter the rocks with which they come in contact. When this alteration has proceeded so far as to induce a crystalline or semi-crystalline character, the rock so altered is said to be metamorphosed. Metamorphism has likewise been produced by the chemical action of percolating water, which frequently dissolves out certain minerals, and replaces these with others having often a very different chemical composition. But metamorphism on the large scale--that is to say, metamorphism which has affected wide areas, such as the northern Highlands of Scotland and wide regions in Scandinavia, or the still vaster areas in North America--has most probably been effected both by the agency of heat and chemical action, at considerable depths, and under great pressure. When we observe what effect can be produced by heat upon rocks, under little or no pressure, and how water percolating from above gradually changes the composition of some rock-masses, we may readily believe that at great depths, where the heat is excessive, such metamorphic action must often be intensified. Thus, for example, limestone heated in the usual way gives off its carbonic acid gas, and is reduced to quicklime; but, under sufficient pressure, this gas is not evolved, the limestone becoming converted into a crystalline marble. Some crystalline limestones, indeed, have all the appearance of having at one time been actually melted and squirted under great pressure into seams and cracks of the surrounding strata. Heated water would appear to have been the agent to which much of the metamorphism which affects the rocky strata must be attributed. But the mode or modes in which it has acted are still somewhat obscure; as may be readily understood when it is remembered how difficult, and often how impossible it is to realise or reproduce in our laboratories the conditions under which deep-seated metamorphic action must frequently have taken place. In foliated rocks, the minerals are chiefly quartz, felspar, and mica, talc, or chlorite. The ingredients of these minerals undoubtedly existed in a diffused state in the original rocks, and heated water charged with alkaline carbonates, as it percolated through the strata, either along the layers of bedding or lines of cleavage, slowly acted upon these, dissolving and redepositing them, and thus inducing segregation.

There is every kind of gradation in metamorphism. Thus, we find certain rocks which are but slightly altered--their original character being still quite apparent; while, in other cases, the original character is so entirely effaced that we can only conjecture what that may have been. When we have a considerable thickness of metamorphic rocks which still exhibit more or less distinct traces of bedding, like the successive beds of gneiss, mica-schist, and quartz rock of the Scottish Highlands, we can hardly doubt that the now crystalline masses are merely highly altered aqueous strata. But there are cases where even the bedding becomes obliterated, and it is then much more difficult to determine the origin of the rocks. Thus, we find bedded gneiss passes often, by insensible gradations, into true amorphous granite. There has been much difference of opinion as to the origin of granite--some holding it to be an igneous rock, others maintaining its metamorphic origin. It is probably both igneous and metamorphic, however. If we conceive of certain aqueous rocks becoming metamorphosed into gneiss, we may surely conceive of the metamorphism being still further continued until the mass is reduced to a semi-fluid or pasty condition, when all trace of foliation and bedding might readily disappear, and the weight of the superincumbent strata would be sufficient to force portions of the softened mass into cracks and crevices of the still solid rocks above and around it. Hence we might expect to find the same mass of granite passing gradually in some places into gneiss, and in other places protruding as veins and dykes into the surrounding rocks; and this is precisely what occurs in nature.

84. Mineral veins have, as a rule, been formed by water depositing along the walls of fissures the various matters which they held in solution, but certain kinds of veins (such as quartz veins in granite) probably owe their origin to chemical action which has induced the quartz to segregate from the rock mass. Some have maintained that the metallic substances met with in many veins owe their deposition to the action of currents of voltaic electricity; while others have attributed their presence to sublimation from below, the metals having been deposited in the fissures very much as lead is deposited in the chimney of a leadmill. But in many cases there seems little reason to doubt that the ores have merely been extracted from the rocks, and re-deposited in fissures, by

water, in the same way as the other minerals with which they are associated.

PHYSIOGRAPHY.

85. Denudation.--By the combined action of all the geological agencies which have been described in the preceding sections, the earth has acquired its present diversified surface. Valleys, lacustrine hollows, table-lands, and mountains have all been more or less slowly formed by the forces which we see even now at work in the world around us. When we reflect upon the fact that all the inclined strata which crop out at the surface of the ground are but the truncated portions of beds that were once continuous, and formed complete anticlinal arches or curves, we must be impressed with the degree of denudation, or wearing-away, which the solid strata have experienced. If we protract in imagination the outcrop of a given set of strata, we shall find them curving upwards into the air to a height of, it may be, hundreds or even thousands of feet, before they roll over to come down and fit on to the truncated ends of the beds on the further side of the anticline (see figs. 9 and 11, pages 33, 34). Dislocations or faults afford further striking evidence in the same direction. Sometimes these have displaced the strata for hundreds and even thousands of feet--that is to say, that a bed occurring at, for example, a few feet from the surface upon one side of a fault, has sunk hundreds or thousands of feet on the other side. Yet it often happens that there is no irregularity at the surface to betray the existence of a dislocation. The ground may be flat as a bowling-green, and yet, owing to some great fault, the rocks underneath one end of the flat may be geologically many hundred feet, or even yards, higher or lower than the strata underneath the other end of the same level space. What has become of the missing strata? They have been carried away grain by grain by the denuding forces--by weathering, rain, frost, and fluviatile and marine action. The whole surface of a country is exposed to the abrading action of the subaerial forces, and has been carved by them into hills and valleys, the position of which depends partly upon the geological structure of the country, and partly upon the texture and composition of the rocks. The original slope of the surface, when it was first elevated out of the sea, would be determined by the action of the subterraneous forces--the dominant parts,

whether table-lands or undulating ridges, forming the centres from which the waters would begin to flow. After the land had been subjected for many long ages to the wearing action of the denuding agents, it is evident that the softer rocks--those which were least capable of withstanding weathering and erosion--would be more worn away than the less easily decomposed masses. The latter would, therefore, tend to form elevations, and the former hollows. This is precisely what we find in nature. The great majority of isolated hills and hilly tracts owe their existence as such merely to the fact that they are formed of more durable materials than the rock-masses by which they are surrounded. When a line of dislocation is visible at the surface, it is simply because rocks of unequal durability have been brought into juxtaposition. The more easily denuded strata have wasted away to a greater extent than the tougher masses on the other side of the dislocation. Nearly all elevations, therefore, may be looked upon as monuments of the denudation of the land; they form hills for the simple reason that they have been better able to withstand the attacks of the denuding agents than the rocks out of which the hollows have been eroded.

86. To this general rule there are exceptions, the most obvious being hills and mountains of volcanic origin, such as Hecla, Etna, Vesuvius, &c., and, on a larger scale, the rocky ridge of the Andes. Again, it is evident that the great mountain-chains of the world are due in the first place to upheaval; but these mountains, as we now see them--peaks, cliffs, precipices, gorges, ravines--have been carved out of the solid block, as it were, by the ceaseless action of the subaerial forces. The direction of river-valleys has in like manner been determined in the first place by the original slope of the land; but the deep dells, the broad valleys and straths, have all been scooped out by running water. The northern Highlands of Scotland, for example, evidently formed at one time a broad table-land, elevated above the level of the sea by the subterranean forces. Out of this old table-land the denuding agents, acting through untold ages, have carved out all the numerous ravines, glens, and valleys, the intervening ridges left behind now forming the mountains. It is true that now and again streams are found flowing in the direction of a fault, but that is simply because the dislocation is a line of weakness, along which it is easier for the denuding forces to act. For one fault that we find running

parallel to the course of a river, we may observe hundreds cutting across its course at all angles. The great rocky basins occupied by lakes, which are so abundant in the mountainous districts of temperate regions and in northern latitudes, are believed to have been excavated by the erosive power of glacier-ice; and they point, therefore, to a time when our hemisphere must have been subjected to a climate severe enough to nourish massive glaciers in the British Islands and similar latitudes. It may be concluded that the present physiography of the land is proximately due solely to the action of the denuding agents--rain, frost, rivers, and the sea. But the lines along which these agents act with greatest intensity have been determined in the first place by the subterranean forces which upheaved the solid crust into great table-lands or mountain undulations. Both the remote and the proximate causes of the earth's surface-features, however, have acted in concert and contemporaneously, for no sooner would new land emerge above the sea-level than the breakers would assail it, and all the forces of the atmosphere would be brought to bear upon it--rain, frost, and rivers--so that the beginning of the sculpturing of hill and valley dates back to the period when the present lands were slowly emerging from the ocean. So great is the denudation of the land, that in process of time the whole would be planed down to the level of the sea, if it were not for the subterranean forces, which from time to time depress and elevate different portions of the earth's crust. It can be proved that strata miles in thickness have been removed bodily from the surface of our own country by the seemingly feeble agents of denudation. All the denuded material--mud, sand, and gravel--carried down into the sea has been re-arranged into new beds, and these have ever and anon been pushed up to the light of day, and scarped and channelled by the denuding forces, the resulting detritus being swept down as before into the sea, to form fresh deposits, and so on. It follows, therefore, that the present arrangement of land and sea has not always existed. There was a time before the present distribution of land obtained, and a time will yet arrive when, after infinite modifications of surface and level, the continents and islands may be entirely re-arranged, the sea replacing the land, and vice versa. To trace the history of such changes in the past is one of the great aims of the scientific geologist.

PALAEONTOLOGY.[F]

[F] Palaios, ancient, onta, beings, and logos, a discourse.

87. Fossils.--In our description of rock-masses, and again in our account of geological agencies, we referred to the fact that certain rocks are composed in large measure, or exclusively, of animal or vegetable organisms, or of both together; and we saw that analogous organic formations were being accumulated at the present time. But we have deferred to this place any special account of the organic remains which are entombed in rocks. Fossils, as these are called, consist generally of the harder and more durable parts of animals and plants, such as bones, shells, teeth, seeds, bark, and ligneous tissues, &c. But it is usual to extend the term fossil to even the casts or impressions of such remains, and to foot-marks and tracks, whether of vertebrates, molluscs, crustaceans, or annelids. The organic remains met with in the rocks have usually undergone some chemical change. They have become petrified wholly or in part. The gelatine which originally gave flexibility to some of them has disappeared, and even the carbonate and phosphate of lime of the harder parts have frequently been replaced by other mineral matter, by flint, pyrites, or the like. So perfect is the petrifaction in many cases, that the most minute structures have been entirely preserved--the original matter having been replaced atom by atom. As a rule, fossils occur most abundantly and in the best state in clay-rocks, like shale; while in porous rocks, like sandstone, they are generally poorly preserved, and not of so frequent occurrence. One reason for this is, that clay-rocks are much less pervious than sandstone, and their imbedded fossils have consequently escaped in greater measure the solvent powers of percolating water. But there are other reasons for the comparative paucity of fossils in arenaceous strata, as we shall see presently.

88. Proofs of varied Physical Conditions.--Organic remains are either of terrestrial, fresh-water, or marine origin, and they are therefore of the utmost value to the geologist in deciphering the history of those great changes which have culminated in the present. But we can go a step further than this. We know that at the present day the distribution of animal and vegetable life is due

to a variety of causes--to climatic and physical conditions. The creatures inhabiting arctic and temperate regions contrast strongly with those that tenant the tropics. So also we observe a change in animal and vegetable forms as we ascend from the low grounds of a country to its mountain heights. Similar changes take place in the sea. The animals and plants of littoral regions differ from those whose habitat is in deeper water. Now, the fossiliferous strata of our globe afford similar proofs of varying climatic and physical conditions. There are littoral deposits and deep-sea accumulations: the former are generally coarse-grained (conglomerates, grit, and sandstone); the latter are for the most part finer-grained (clay, shale, limestone, chalk, &c.); and both inshore and deep-water formations have each their peculiar organic remains. Again, we know that some parts of the sea-bottom are not so prolific in life as others--where, for example, any considerable deposit of sand is taking place, or where sediment is being constantly washed to and fro upon the bottom, shells and other creatures do not appear in such numbers as where there is less commotion, and a finer and more equable deposit is taking place. It is partly for the same reason that certain rocks are more barren of organic remains than others.

89. Fossil Genera and Species frequently extinct.--It might perhaps at first be supposed that similar rocks would contain similar fossils. For example, we might expect that formations resembling in their origin those which are now forming in our coral seas would also, like the latter, contain corals in abundance, with some commingling of shells, crustaceans, fish, &c., such as are peculiar to the warm seas in which corals flourish. And this in some measure holds good. But when we examined carefully the fossils in certain of the limestones of our own country, we should find that while the same great orders and classes were actually present, yet the genera and species were frequently entirely different; and not only so, but that often none of these were now living on the earth. Moreover, if we extended our research, we should soon discover that similar wide differences actually obtained between many of the limestones themselves and other fossiliferous strata of our country.

90. Fossiliferous Strata of Different Ages.--Another fact would also gradually

dawn upon us--this, namely, that in certain rocks the fossils depart much more widely from analogous living forms, than the organic remains in certain other rocks do. The cause of this lies in the fact that the fossiliferous strata are of different ages; they have not all been formed at approximately the same time. On the contrary, they have been slowly amassed, as we have seen, during a long succession of eras. While they have been accumulating, great vicissitudes in the distribution of land and sea have taken place, climates have frequently altered, and the whole organic life of the globe has slowly changed again and again--successive races of plants and animals flourishing each for its allotted period, and then becoming extinct for ever.[G] Thus, strata formed at approximately the same time contain generally the same fossils; while, on the other hand, sedimentary deposits accumulated at different periods are charged with different fossils. Fossils in this way become invaluable to the geologist. They enable him to identify formations in separate districts, and to assign to them their relative antiquity.[H] If, for example, we have a series of formations, A, B, C, piled one on the top of the other, A being the lowest, and C the highest, and each charged with its own peculiar fossils, we may compare the fossils met with in other sets of strata with the organic remains found in A, B, C. Should the former be found to correspond with the fossil contents of B, we conclude that the rocks in which they occur are approximately of contemporaneous origin with B, even although the equivalents of the formations A and C should be entirely wanting. Further, we soon learn that the order of the series A, B, C, is never inverted. If A be the lowest, and C the highest stratum in one place, it is quite certain that the same order of succession will obtain wherever the equivalents of these strata happen to occur together. But the succession of strata is not invariably the same all the world over; in some countries, we may have dozens of separate formations piled one on the top of the other; in other countries, many members of the series are absent; in brief, blanks in the succession are of constant occurrence. But by dovetailing, as it were, all the formations known to us, we are enabled to form a more or less complete series of rocks arranged in the order of their age. A little reflection will serve to shew that the partial mode in which the strata are distributed over the globe arises chiefly from two causes. We have to remember, first, that the deposits themselves were laid down only here and there in irregular spreads and patches--opposite the mouths of rivers, at various points along the ancient coast-lines, and over certain areas in the deeper abysses of the ocean--the coarser accumulations being of much less extent than those formed of finer materials. And, second, we must not forget the intense denudation which they have experienced, so that miles and miles of strata which once existed have been swept away, and their materials built up into new formations.

[G] To this there are some exceptions. Certain small foraminifers, for example, met with in some of the oldest formations, do not seem to differ from species which are still living. The genus Lingula (Mollusca) has also come down from remotest ages, having outlived all its earlier associates.

[H] This holds strictly true, however, only in regard to comparatively limited areas. The student must remember that strata occurring in widely separate regions of the earth, even although they contain very much the same assemblage of fossils, are not necessarily contemporaneous, in the strict meaning of the word; for the fauna and flora (the animal and plant life) may have died out, and become replaced by new forms more rapidly in one place than another. The term 'contemporaneous,' therefore, is a very lax one, and may sometimes group together deposits which, for aught that we can tell, may really have been accumulated at widely separated times.

91. Gradual Extinction of Species.--When a sufficient number of fossils has been diligently compared, we discover that those in the younger strata approach most nearly to the present living forms, and that the older the strata are, the more widely do their organic remains depart from existing types of animals and plants. We may notice also, that when a series of beds graduate up into each other, so that no strongly marked line separates the overlying from the underlying strata, there is also a similar gradation amongst the fossils. The fossils in the highest beds may differ entirely from those in the lowest; but in the middle beds there is an intermingling of forms. In short, it is evident that the creatures gradually became extinct, and were just as gradually replaced by new forms, until a time came when all the species that were living while the

lowest beds were being amassed, at last died out, and a complete change was effected.

92. Proofs of Cosmical Changes of Climate.--From the preceding remarks it will be also apparent that fossils teach us much regarding the climatology of past ages. They tell us how the area of the British Islands has experienced many vicissitudes of climate, sometimes rejoicing in a warm or almost tropical temperature, at other times visited with a climate as severe as is now experienced in arctic and antarctic regions. Not only so, but we learn from fossils that Greenland once supported myrtles and other plants which are now only found growing under mild and genial climatic conditions; while, on the other hand, remains of arctic mammals are met with in the south of France. Such great changes of climate are due, according to Mr Croll, to variations in the eccentricity of the earth's orbit combined with the precession of the equinox. It is well known that the orbit of our earth becomes much more elliptical at certain irregularly recurring periods than it is at present. During a period of extreme ellipticity, the earth is, of course, much further away from the sun in aphelion[I] than it is at a time of moderate ellipticity, while, in perihelion,[J] it is considerably nearer. Now, let us suppose that, at a time when the ellipticity is great, the movement known as the precession of the equinox has changed the incidence of our seasons, so that our summer happens in perihelion and not in aphelion, while that of the southern hemisphere occurs in aphelion, and not, as at present, in perihelion. Under such conditions, the climate of the globe would experience a complete change. In the northern hemisphere, so long and intensely cold would the winter be, that all the moisture that fell would fall as rain, and although the summer would be very warm, it would nevertheless be very short, and the heat then received would be insufficient to melt the snow and ice which had accumulated during the winter. Thus gradually snow and ice would cover all the lands down to temperate latitudes. In the southern hemisphere, the reverse of all this would obtain. The winter there would be short and mild, and the summer, although cool, would be very long. But such changes would bring into action a whole series of physical agencies, every one of which would tend still further to increase the difference between the climates of the two hemispheres. Owing to the vast

accumulation of snow and ice in the northern hemisphere, the difference of temperature between equatorial and temperate and polar regions would be greater in that hemisphere than in the southern. Hence the winds blowing from the north would be more powerful than those coming from the southern and warmer hemisphere, and consequently the warm water of the tropics would necessarily be impelled into the southern ocean. This would tend still further to lower the temperature of our hemisphere, while, at the same time, it would raise correspondingly the temperature at our antipodes. The general result would be, that in our hemisphere ice and snow would cover the ground down to low temperate latitudes--the British Islands being completely smothered under a great sea of confluent glaciers. In the southern hemisphere, on the contrary, a kind of perennial summer would reign even up to the pole. Such conditions would last for some ten or twelve thousand years, and then, owing to the precession of the equinox, a complete change would come about--the ice-cap would disappear from the north, and be replaced by continuous summer, while at the same time an excessively severe or glacial climate would characterise the south; and such great changes would occur several times during each prolonged epoch of great eccentricity. This, in few words, is an outline of Mr Croll's theory. That theory is at present sub judice, but there can be no doubt that it gives a reasonable explanation of many geological facts which have hitherto been inexplicable. Of course, it is not maintained that all changes of climate are due directly or indirectly to astronomical causes. Local changes of climate--changes affecting limited regions--may be induced by mutations of land and sea, resulting in the partial deflection of ocean currents, which are the chief secondary means employed by nature for the distribution of heat over the globe's surface.

[I] Apo, away from; helios, the sun.

[J] Peri, round about or near by; helios, the sun.

From what has been stated in the foregoing paragraphs, it is clear that in our endeavours to decipher the geological history of our planet, palaeontological must go hand in hand with stratigraphical evidence. We may indeed learn much from the mode of arrangement of the rocks themselves. But the test of superposition does not always avail us. It is often hard, and sometimes quite impossible, to tell from stratigraphical evidence which are the older rocks of a district. In the absence of fossils we must frequently be in doubt. But physical evidence alone will often afford us much and varied information. It will shew us what was land and what sea at some former period; it will indicate to us the sites of ancient igneous action; it will tell us of rivers, and lakes, and seas which have long since passed away. Nay, in some cases, it will even convince us that certain great climatic changes have taken place, by pointing out to us the markings, and debris, and wandered blocks which are the sure traces of ice action, whether of glaciers or icebergs. The results obtained by combining physical and palaeontological evidence form what is termed Historical Geology.

HISTORICAL GEOLOGY.

93. The fossiliferous strata, as they are generally termed, have been chronologically arranged in a series of formations, each of which is characterised by its own peculiar suites of fossils. Their relative age has been determined, as we have indicated above, by their fossils, and also by certain physical tests, the chief of these being superposition. It holds invariably true that a formation, A, found resting upon another series of strata, B, will always occur in precisely the same position, wherever these two deposits occur together. If B should appear in some place as resting upon A, we may be sure that the beds have been inverted during the contortion of the strata consequent upon subterranean action (see fig. 11, page 34). Again, another useful test of the relative age of strata lies in the circumstance that one is often made up or contains fragments of the other. In this case, then, it is quite clear which is the more recent accumulation. These tests have now been applied to the strata in many parts of the world, and the result is that geologists have been able to arrive at a chronological arrangement or classification, and so to construct a table shewing the relative position which would be occupied by all the different formations, if these occurred together in one place. In the British Islands the long series of strata is well developed, but many of the formations

are much more meagrely represented than their equivalents in other countries. But even when we attempt to fill up the blanks in our own series by dovetailing with them the strata of foreign countries, there yet remain numerous breaks in the succession, pointing to the fact that the stony record is a very fragmentary one at the best. No doubt there are many large tracts of the earth's surface which have not yet been investigated, and when these are known we may hope to have our knowledge greatly increased. But no one who reflects upon the mode of origin of the fossiliferous strata, and the wonderful mutations which the earth has undergone, can reasonably anticipate that a perfect and complete record of the geological history of our planet shall ever be compiled from the broken and fragmentary testimony of the rocks.

94. The following table gives the names of the different formations arranged in the order of their superposition, the youngest being at the top, and the oldest known at the bottom:

IV. POST-TERTIARY OR QUATERNARY-- Historical or Recent. Pleistocene.

III. TERTIARY OR CAINOZOIC-- Pliocene. Miocene. Eocene.

II. SECONDARY OR MESOZOIC-- Cretaceous. Jurassic. Triassic.

I. PRIMARY OR PALAEOZOIC-- Permian. Carboniferous. Devonian and Old Red Sandstone. Silurian. Cambrian. Laurentian or Pre-Cambrian.

95. The PRIMARY formations are so called because they are the oldest known to us: they are not necessarily the first-formed aqueous deposits. Dr Hutton said truly: There is no trace of a beginning, and no signs of an end. In the PRIMARY or PALAEOZOIC (ancient-life) formations are found the earliest traces of life. The forms as a rule depart very widely from those with which we are acquainted now. The Laurentian rocks have yielded only one fossil--a large foraminifer named Eozoon Canadense. The Cambrian formation contains but few fossils--crustaceans, molluscs, zoophytes, and worm-tracks.

The Silurian strata are often abundantly fossiliferous. All the great classes of invertebrates are represented, and fish remains also occur. The Devonian and Old Red Sandstone are also characterised by the presence of an abundant fauna. In the Old Red Sandstone are numerous fish remains; it appears to have been an estuarine or lacustrine deposit; the Devonian, on the other hand, was marine, like the Silurian and Cambrian. The Carboniferous formation is the chief repository of coal in Britain. It consists of terrestrial, fresh or brackish water, and marine deposits. The fauna and flora of the Permian, which is partly a marine and partly a fresh-water formation, are allied, upon the whole, to those of the Carboniferous, but offer at the same time many contrasts.

96. The SECONDARY OR MESOZOIC (middle-life) formations contain assemblages of fossils which do not depart so widely from analogous living forms as those belonging to Palaeozoic times. The Triassic strata yield abundance of rock-salt. In Britain they contain very few fossils, but these are more abundant in the Triassic deposits of foreign countries. The oldest known mammals first appear in this formation. The Jurassic formation is very highly fossiliferous. It is distinguished by the occurrence of numerous reptilian remains. Nearly all the beds of this formation are marine, but there are associated with these the remains of a forest or old land surface, and a considerable accumulation of estuarine or fresh-water deposits; impure coals also occur in this formation. The Cretaceous strata are almost wholly marine, and chiefly of deep-water origin. But some land-plants are found, chiefly ferns, conifers, and cycads. Near the base of the formation occurs a great river deposit (Weald clay) with numerous remains of reptiles.

97. Among the oldest strata of the TERTIARY or CAINOZOIC (recent-life) division we meet with the dawn of the existing state of the testaceous fauna-- the Eocene (eos, dawn, and kainos, recent) containing three and a half per cent. of recent species among its shells. The proportion of recent species increases in the Miocene (meion, less, and kainos, recent), although the majority of the molluscs entombed in that formation belong to extinct species. In the Pliocene (pleion, more, and kainos, recent), however, the extinct species are in a minority.

The POST-TERTIARY or QUATERNARY division comprises the concluding chapters of geological history. The Pleistocene (pleistos, most, and kainos, recent) contains no extinct species of shells, but a number of extinct mammalia. In the Recent deposits all the species of animals and plants are living. The Tertiary and Quaternary formations are partly of marine and partly of terrestrial and fresh-water origin. At the close of the Tertiary period the 'glacial epoch' of Pleistocene times began, and the British Islands and a large part of northern Europe and North America were then cased in snow and ice. Traces of glacial conditions have also been met with in the Eocene and Miocene. The evidence furnished by Palaeozoic and Mesozoic formations points chiefly to mild, genial, and sometimes tropical conditions. But traces of ice action are occasionally noted (namely, in the Silurian, Old Red Sandstone, Carboniferous, Permian, and Cretaceous formations), pointing, perhaps, in some of the cases, to former alternations of cold and warm periods. Indeed, the belief is now gaining ground, that the so-called glacial epoch of Pleistocene times was not one long continuous age of ice, but rather consisted of an alternation of warm and cold periods. And it is not improbable, but highly likely, that similar alternations of climate have happened during every period of great eccentricity of the earth's orbit.

QUESTIONS.

Section 1. What is Geology?

2. Define the term rock. How many classes of rock are there?

3, 4, 5. Into what groups are the mechanically formed rocks divided? Define the terms conglomerate, sandstone, and shale.

6. What is the nature of the rocks belonging to the Aerial or Eolian group?

7. Give an example of a chemically formed rock.

8. Give examples of organically derived rocks.

9. What kinds of rocks are embraced by the Metamorphic class?

10. What are igneous rocks?

12. What is the mineralogical composition of granite?

13. What is meant by a mineral?

14. Name five minerals which do not contain oxygen. Where does fluor-spar occur? What is the element that enters most largely into the composition of the earth's crust?

15. Name the forms under which the mineral quartz occurs. Name some of the oxides of iron. What is iron pyrites?

16. Name two sulphates. Name two carbonates. Name some of the silicates. In what kinds of rock is augite found? Where does it never occur? In what kinds of rock does hornblende usually occur? Mention three species of felspar. What is one of the most distinguishing characteristics of mica? Name three silicates of magnesia. Mention some of their distinguishing peculiarities. Where do zeolites commonly occur?

17. What is a quartzose conglomerate? What is a calcareous conglomerate?

18. What is grit? What is freestone? To what are the various colours of sandstone due? What is shale?

19. Name some typical Eolian rocks, and tell where they occur.

20. How do stalactites and stalagmites occur? What is siliceous sinter, and how does it occur? How does rock-salt occur?

21. Mention some of the varieties of limestone. What is cornstone? What is the composition of dolomite?

22. Name some of the varieties of coal.

23. What is quartzite?

24. Describe clay-slate.

25. Mention some altered limestones.

26. What are schists? Name and give the mineralogical composition of three schists.

27. What is the general character of metamorphic rocks?

28. How would you classify granite?

29. What is the mineralogical composition of syenite and diorite?

30. How do we distinguish the two groups into which igneous rocks are subdivided? What is meant by the terms amygdaloidal and porphyritic?

31. Name some rocks that belong to the acidic group. What is quartz-porphyry?

32. Give examples of augitic igneous rocks. Name a hornblendic igneous rock.

33. What are fragmental igneous rocks? What is the difference between trappean breccia and trappean conglomerate?

34. What is meant by the terms stratum, strata, and stratified? What is the difference between lamination and bedding? What is a section?

35. What is false bedding?

36. Briefly describe the general appearance of mud-cracks and rain-prints, and say how these have been formed.

37. What is meant by a succession of strata?

38. Which kinds of stratified rocks generally have the greatest extension?

39. How do beds terminate?

40. How may planes of bedding sometimes indicate a break in the succession of strata?

41. What is the nature of joints? What are master-joints, and what is their probable cause?

42. What is cleavage, and what is its effect upon the bedding of rocks?

43. What is foliation?

44. Give examples of concretionary rocks. What is the nature of chert and flint nodules?

45. Define the terms dip and strike. What is the crop of a bed? What are anticlines and synclines?

46. What is meant by an inversion of strata?

47. How does contemporaneous erosion indicate a pause in the deposition of a series of strata?

48. What is meant by unconformability? How does unconformability prove a

lapse of time between the accumulation of the underlying and overlying strata?

49. What is overlap?

50. What is a fault? What is hade? How are the strata affected on either side of a fault? What is the appearance called slickensides? Under what circumstances should we term a fault a downthrow? and when should we term it an upcast? How is the approximate age of a fault sometimes shewn?

51. What are metamorphic rocks, and what is their general appearance? In what districts of the British Islands are they most abundantly developed? What are some of the appearances relied upon for distinguishing metamorphic from igneous granite?

52. How do igneous rocks occur? Define what is meant by contemporaneous and subsequent or intrusive igneous rocks. How does a contemporaneous igneous rock affect the beds upon which it rests? What is the character of the bed overlying a contemporaneous rock? What is the general structure of a contemporaneous igneous rock? What is meant by vesicular structure? What is the general texture of a contemporaneous igneous rock? What is the nature of the jointing in igneous rocks? What is wacke?

53. What is the nature of the beds of breccia, conglomerate, ash, and tuff, with which contemporaneous igneous rocks are often associated? What is a neck of volcanic agglomerate? How are the strata affected at their junction with a 'neck'?

54. How do intrusive igneous rocks occur? How do intrusive sheets occur? What effect have they produced upon the strata above and below them? What is a dyke? What relation do they occasionally bear to sheets of igneous rock? What is a neck of intrusive igneous rock, and how have the strata surrounding it been affected?

55. Mention some of the contrasts between intrusive and contemporaneous

igneous rocks. What alteration is produced upon coal with which an intrusive sheet has come in contact?

56. What are mineral veins? What is the nature of the quartz veins in granite? How are the minerals usually arranged in the great metalliferous veins? What is a pipe-vein?

57. What are the great geological agents of change?

58. What is meant by weathering? How are rocks affected at the surface in tropical countries? What chemical effect has the atmosphere on calcareous rocks? How is soil formed? How are sand dunes formed? Mention some effects of the transporting power of the atmosphere.

59. Mention some of the chemical effects of interstitial water. What is the origin of travertine or calcareous tufa?

60. How have stalactites and stalagmites been formed? Give some instances of the solvent power of springs.

61. How are caves in limestone formed? Describe some of the appearances of a country composed of calcareous rocks. Describe briefly how a river erodes its channel.

62. Describe the geological action of rain.

63. What do chemical analyses of river-water prove? Give an example. What are pot-holes? Give an example of the erosive power of running water. What amount of mud is carried in suspension by the Mississippi, and discharged annually into the sea? What estimate has been formed of the total amount of mineral matter annually transported by that river?

64. What is alluvium? How is it formed? and mention some examples of its occurrence.

65. How is sediment deposited by a river in a lake?

66. What is the difference between lacustrine and fluvio-marine deposits? What is a delta?

67. Describe the geological action of frost. Describe the geological action of river-ice.

68. What are glaciers? What thickness do they attain in the Alps? What is their rate of motion? What are crevasses, and how do they originate? What are superficial moraines? What are terminal moraines? What changes does a glacier effect upon its bed, and how are these modifications produced? What is the character of a glacial river? What is the origin of icebergs? How is the general absence of blocks and stones in Greenland icebergs to be explained? What is the nature of a submarine terminal moraine? What is the ice-foot? What is the chief agent in distributing erratic stones and blocks over the sea-bottom? What effect upon the sea-bed must stranding icebergs produce?

69. What are some of the chemical compounds held in solution in sea-water? Which of these go to form the shells and skeletons of marine animals?

70. Describe the action of breakers on a sea-coast. How does frost aid the wasting action of breakers? What effect has the nature of the rocks in the production of inequalities in a coast-line? Upon what part of the sea-bottom does the material derived by the action of the breakers chiefly accumulate? What effect have the tides and ocean currents in the distribution of sediment?

71. What is the general rule as regards fine-grained and coarse-grained deposits? Mention a partial exception to this rule. What effect have tidal currents in shallow seas?

72. How are rocks disintegrated through the action of plants? What is peat? What may be inferred from the occurrence of shell-marl underneath peat?

What does the appearance of roots and trunks of trees, and of remains of land animals under peat, indicate?

73. What, generally, is the geological action of animal life?

74. What is coral? What is a fringing reef? What is the general character of a barrier reef? Give an example of one. What is an atoll? What is the nature of coral rock? What is Mr Darwin's theory of the formation of coral reefs?

75. What is the nature of the Atlantic ooze? In what respects may it eventually come to resemble chalk and limestone? Mention an instance of the abundant occurrence in the sea of animalcules with siliceous coverings and skeletons. What is the nature of the red clay found at great depths in the Atlantic and Southern Oceans?

76. What are some of the notions held in regard to the internal condition of the earth? At what (average) rate does the temperature of the earth's crust increase as we descend from the surface?

77. What is the nature of the movements to which the earth's crust is subjected?

78. Describe the hypotheses advanced to account for earthquakes. Mention some of the effects of earthquakes--1st, as regards alterations of level; and 2d, as regards modifications of the surface.

79. Mention a good example of tranquil elevation and depression of the earth's crust. Mention some of the proofs of an elevatory movement. Give proofs that shew depression of the land. How may certain former changes of sea-level be accounted for without inferring any movement of the land?

80. What effect must depression have upon the strata forming the earth's crust? What is the result of a movement of elevation? What is the cause of cleavage?

81. What is the nature of the materials thrown out during volcanic eruptions? What is the general structure of a volcanic cone? How does molten rock make its escape from the orifice of eruption? What is the meaning of the terms lapillo, puzzolana, and ceneri?

82. What is lava? Describe the general appearance and mode of progression of a stream of lava. What effect is produced upon fragments of rock caught up and inclosed in lava; and what changes are caused in the pavement upon which it cools? How does a lava stream entering a lake or the sea behave in regard to the sediment gathering therein? To what is the basaltic structure due? How are the axes of the prisms in a columnar igneous rock arranged? Name some of the varieties of lava. What is the origin of the vesicular structure in igneous rocks? What portions of a bed of lava are most frequently scoriaceous? In what kinds of lava is the vesicular structure most abundantly met with? How have the vesicles become flattened? In what manner have they been filled with mineral matter? What is the origin of the dykes of modern volcanic districts?

83. How is metamorphism on the large scale supposed to have been induced? How may granite be at one and the same time a metamorphic and igneous rock?

84. Mention some of the views held with regard to the origin of mineral veins.

85. What is denudation? How do inclined strata prove that the strata have been denuded? How do faults afford proof of denudation? What have been the general effects produced by denudation on the face of the land?

86. What part have the subterranean forces acted in the formation of mountains? To what geological action is the present aspect of these mountains due? What has determined the direction of river valleys? How have the valleys, dells, &c. been formed? What effect have faults had in determining the direction of river valleys? What is supposed to be the origin of the deep rock-basins occupied by many fresh-water lakes? How is the waste of land by denudation compensated?

87. What are fossils? What is meant by petrifaction? In what kind of rocks do fossils occur most abundantly, and in the best state of preservation? and what reason can be given for this?

88. How do fossils afford proof of varied physical conditions? Give a reason for some rocks being more barren of fossils than others.

89. State some of the characters which distinguish broadly the older fossiliferous strata from those similar accumulations which are being formed in our own day.

90. How may we identify formations in separate districts? How is the interrupted and partial distribution of strata to be accounted for?

91. In what respect do the fossils in younger strata differ from those in older strata? What general proof can be adduced to shew that species have become gradually extinct?

92. Give an instance how fossils prove changes of climate in the past. What is supposed to be the cause of great cosmical changes of climate? Describe Mr Croll's theory of cosmical changes of climate.

93. What is the test of superposition? Mention another test of the relative age of strata. 94. Name the four great divisions under which the fossiliferous rocks are arranged.

95. Name the Primary or Palaeozoic formations. What are the principal kinds of fossils found in the Old Red Sandstone? Which formation is the chief repository of coal in Britain?

96. In what other formations do coals occur? In which formation do the oldest known mammals occur? Name the Secondary formations.

97. Name the Tertiary formations. What kind of climate characterised the

northern hemisphere at the beginning of Pleistocene times? What kinds of climate would appear from the evidence to have chiefly prevailed in Primary, Secondary, and Tertiary ages? Have we any trace of frigid conditions during these ages? What is the growing opinion with regard to the climatic conditions during the glacial period of Pleistocene times?

THE END

www.ingramcontent.com/pod-product-compliance
Lightning Source LLC
Chambersburg PA
CBHW051815170526
45167CB00005B/2022